The Galloping Sausage and Other Train Curiosities

150 Steam Railway Events and Stories

Geoff Body and Ian Body

Front cover images: *Above*: 10000 Doncaster, 'Galloping Sausage'. (With thanks to John Chalcraft at Rail Photoprints); *Below*: *Flying Scotsman*. (G. Body)
Back cover images: *Left*: *Earl of Merioneth*. (G. Body); *Right*: Level crossing at Halesworth Station. (G. Body)

First published 2016

The History Press
The Mill, Brimscombe Port
Stroud, Gloucestershire, GL5 2QG
www.thehistorypress.co.uk

British Library Cataloguing in Publication Data.
A catalogue record for this book is available from the British Library.

ISBN 978 0 7509 6593 4

Typesetting and origination by The History Press
Printed in Turkey

CONTENTS

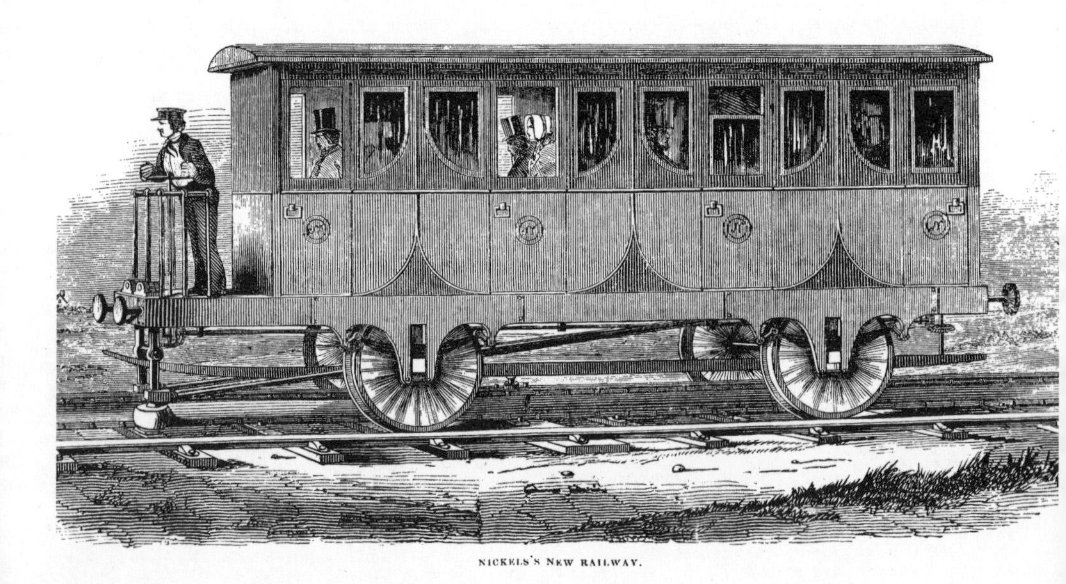

Nickel's railway. (*Illustrated London News*)

INTRODUCTION, SOURCES AND ACKNOWLEDGMENTS

It is doubtful whether any strand in the technological development of the railways occurred without deviation from what eventually became the norm. Indeed, imagination and experiment are integral parts of any growth process, and the railway business was no exception. The pioneers starting from scratch had neither precedents to constrain them nor guidelines to fetter their imagination. By later standards many of the methods they tried inevitably attained curiosity status as, in a similar way, did some of the mainstream activity practices which were tried in the search for improvement. Such is the nature of change.

Beginning even before the Rainhill Trials, the course of motive power development not only evolved erratically but wandered considerably in the process. Steam power was always going to replace horses but cable haulage, atmospheric systems and even sail railways were all tried in the quest for superior traction, economy and speed. The same has applied to railway equipment and practices so that there have been no shortage of curious and unusual examples appearing along the way. Nor has there been any lack of out-of-the-ordinary people, both crank and genius.

The aim of this book, following in the path of the earlier *Railway Oddities*, has been to capture a varied but random selection of the curious and the unusual and present the examples in their setting. Hopefully the result might be a readable and possibly astonishing repast of some of the more outstanding of the odd things that have occurred over two centuries of railway history.

The sources consulted have been many and varied, often, like the *Regional Railway History Series* from David & Charles, just giving a clue to something intriguing and worth researching further. The railways' own material, especially the *Great Western Railway Magazine*, the railways' own working instructions and the extensive publicity material, have provided equally rich pickings, as have a whole host of early railway books and documents, local newspapers, libraries and the like.

The authors' own material and experiences are extensive, and many good friends and former colleagues have contributed, especially Bill Parker who has also trawled the goodwill of his own many railway contacts. A number of contributions have also come from library research, aided by people like Linda Tree at Kings Lynn and the Great Yarmouth library staff. Thanks are also due to Jim Dorward and Roy Kethro. We are grateful to all who have made an input.

The illustrations are all from the authors' collections except where otherwise credited.

Great Western Railway steam railcar.

A HOSTILE TAKEOVER

The year 1845 was a very significant one in Norfolk. On 30 July the Norfolk Railway line from Norwich to Brandon was linked with that of the Eastern Counties Railway (ECR) on through Ely and south through Cambridge to London to give Norwich an important railway link with the capital. This development had already raised concerns in Kings Lynn that it would be left out of the exciting new railway age. There were also real fears that the north Norfolk port would lose some of the extensive and lucrative coal traffic brought in from northern harbours for onward movement within East Anglia. The result was intensive local discussions in Kings Lynn resulting in the promotion and incorporation of three railway

Brandon in 1845 was a small and remote place, but important in railway terms as the last link in the creation of the first route from London to Norwich. (*Illustrated London News*)

schemes of its own. These were the Lynn & Ely Railway to connect with the new line to London at Ely, the Lynn & Dereham to provide a route to Norwich and Yarmouth and the Ely & Huntingdon, a truncated version of the original scheme to reach Bedford. All three were amalgamated into the East Anglian Railway (EAR) in 1847.

Work on the East Anglian group's lines was started in 1846 and by the following October the main line had reached Ely and much progress had been made on the other routes as well. Their construction involved little in the way of severe curves or gradients but the company incurred high costs in spanning the waterways of the Fenland sections. Already it was being eyed acquisitively by the growing Eastern Counties Railway as part of an emerging pattern among early railways in which small concerns sought to be acquired at inflated prices and large ones aimed to buy up rivals financially exhausted by construction costs.

This was a period in which the unscrupulous George Hudson was directing the affairs of the Eastern Counties Railway. In negotiations with the East Anglian Railway over a possible lease, he insisted on completion of the latter's Huntingdon and Wisbech lines, knowing full well that they would sap the smaller concern of funds and make it more amenable to acquisition. It certainly did the former and by 1849 the East Anglian Railway was in financial difficulties. It had to sell off land, close stations and reduce its staff levels. A measure of the problem was the decision to provide beds at stations for the use of porters in order to save the cost of night watchmen!

The Eastern Counties Railway had been working the truncated Ely & Huntingdon line between the latter point and St Ives but in 1849 gave notice that it would not continue, leaving the EAR with the problem of how to access this isolated outpost of its network. It did manage to scrape together a few wagons and a tram carriage and find a horse to pull them but then fell foul of the railway commissioners because the horse-drawn 'trains' could not achieve the 12mph required by legislation. In the following year, when the EAR tried to link up with the Great Northern (GNR) at Huntingdon, the contractor for constructing the connecting line insisted on having twenty coaches locked up in a shed as security for his account – such was the Kings Lynn company's parlous financial state.

On 29 June 1850 matters reached their lowest point when possession of the East Anglian Railway was taken by the Official Receiver. Soon various anxious creditors were claiming as security not only the East Anglian's engines but also anything else they could put their hands on. Many locomotives operated with a creditor's representative on the footplate while other assets carried a plate with details of the lien claimant.

Underlying all these events lay the tactics of the EAR's two large neighbours. The Great Northern Railway was anxious to branch out from its new route to the north from Kings Cross and penetrate East Anglia. The Eastern Counties was equally intent on thwarting any such intrusion into what it saw as its territory. It had already offered the East Anglia 25s a day to lease its Huntingdon extremity and now made another derisory offer for the whole undertaking.

Getting to hear of this situation, the local agent for the Great Northern Railway, a Mr Baxter, agreed a more reasonable deal with the East Anglian but his superiors began to have second thoughts when the company tried to run through-trains from Peterborough to Kings Lynn. They found that the ECR had blocked the connection between the two systems and horse buses had to be used to bridge the gap.

Although the East Anglian's creditors had agreed to the end of the receivership when it was leased by the Great Northern, that deal was challenged by a faction of GNR shareholders, especially when the attempt to get an injunction to prevent the Wisbech blockage failed. There were other troubles too. The Norfolk Railway would not let the East Anglian use its Dereham station forcing it to build its own. The cumulative effect was that good sense had to prevail and finally did so when the East Anglian agreed to a better Eastern Counties deal, which came into effect at the beginning of 1852.

DOCTOR DIONYSIUS LARDNER

This complex character was a baffling mixture of intelligence, imagination, flamboyance and self-misdirection. Born in 1793, he attended Trinity College Dublin where he obtained an MA and, after a brief flirtation with the priesthood and a spell in his father's legal business, he began writing books and papers on mathematical and scientific matters. He was appointed professor of Natural Philosophy & Astronomy at University College London in 1828 and remained there for three years during which time he began to take an increasing interest in the steam engine and the emerging railway activity. He then went on to make something of a career out of appearing as an expert in this sphere, finding ready employment with those opposing railway legislation and locking horns with Brunel on several notable occasions. Although a learned man who did much to popularise science, Lardner's largely academic approach led him into some untenable positions, which Brunel's more practical experience easily exposed.

Lardner crossed swords with Brunel in the fifty-seven-day committee hearing of the first Great Western Railway Bill in 1834. He was responsible for many of the questions hurled at Brunel during his seven-day grilling by opposition counsel. Lardner popped up again in the hearings of the following year when he supported those trying to demolish the case for Box Tunnel. According to the professor's figures, if the brakes on a train failed on the down gradient through the tunnel it would emerge at the other end at the terrible speed of 120mph and bring disaster wherever it crashed. Brunel was not slow to point out that the pseudo-expert's calculations had completely ignored both friction and air resistance and that brake failure would not be half the problem that had been suggested.

In 1836 Lardner found time from his prolific writings to give evidence when the London & Brighton Railway scheme was before Parliament. He was back meddling in Great Western Railway (GWR) affairs in 1838, however, when the first section of the company's newly opened railway was having problems with its permanent way and the riding qualities of its carriages. The directors came under pressure from the North Western group of shareholders who, familiar with the pioneer Liverpool & Manchester Railway (L&MR), believed its 4ft 8½in gauge preferable to Brunel's choice of 7ft. The outcome was Brunel having to accept having another engineer's report on the new railway's choice of gauge and its operational troubles, which he did with good grace.

What Brunel could not have been happy about was finding that the reporting team included his old adversary Dionysius Lardner. That the latter should have a special train filled with his own measuring equipment and allowed to roam the system at will must have been an even harder pill to swallow, especially as Brunel had scant regard for the competence of both the man and the devices he was using. The arrangement for Lardner's train to go pretty well where it wanted without prior arrangement was just asking for trouble and this duly came on 26 September 1838 when the 8am regular train ran into Lardner's special and destroyed three of its test carriages. Lardner's reaction is contained in what has been described as 'a very improper letter' to the GWR Board.

Lardner suffered another reverse soon afterwards. The pioneer locomotive *North Star* had been performing badly and Lardner's tests led him to ascribe the loss of tractive effort at higher speeds to the atmospheric resistance to the wide carriages; perhaps he had learned something from the Box Tunnel affair after all! Together Brunel and Daniel Gooch quickly proved him wrong by altering the locomotive's blast pipe and producing a much-improved performance with a lower coke consumption.

Brunel and Lardner were at odds again over Brunel's plans for the SS *Great Western*, the huge steamer which was to capture the transatlantic trade, but by 1840 Lardner had other things to worry about. Having run off to Paris with the wife of an officer in the Dragoon Guards he got a good hiding from the husband and an expensive lawsuit to add to his troubles. With his reputation in Britain shattered, Lardner was to remain in Paris until his death in 1859. He did, however, give the United States the 'benefit' of his expertise for a while before finally lapsing into relative obscurity.

MOMENTS OF PASSION

Despite losing his parents before he was a year old, Charles Blacker Vignoles managed to pursue a varied and successful career, which reached a pinnacle when he was elected president of the Institution of Civil Engineers. Following a short period in the army he spent time in North America working as a land surveyor. He seems to have been largely self-taught from working on a wide variety of projects whilst there.

At the age of 30, and with a wide range of skills, he returned to Britain in 1823. He then worked on the London Docks, surveyed the route of the Liverpool & Manchester Railway and was involved in several other major railway schemes in a number of different countries. Vignoles met John Braithwaite when helping in the preparation of Ericsson's locomotive *Novelty* for the Rainhill Trials, and Braithwaite, in his capacity as chief engineer of the Eastern Counties Railway, was later to involve him in work for that company.

While Vignoles seems to have been a capable man, he was given to temperamental excesses especially when things went wrong. Then he would fly into a tantrum with wild gestures, foot stamping and tearing at his hair. Accompanying this spectacle with a flow of highly colourful oaths he gained a reputation as a man to be avoided when things did not go smoothly.

EDUCATIONAL OPPOSITION

Some nineteenth-century educationalists supported the advent of railways whilst others were implacable in their opposition. None more so than Eton College who opposed both bills for the Great Western Railway despite some strong support for the new company from the town of Windsor. The Eton authorities seemed to fear that the existence of a railway would encourage its pupils to rush up to London to enjoy the temptations offered by the capital. They saw it destroying all the cherished standards of the college even suggesting it might see the traditional English master replaced with the unthinkable, a French mistress, and encouraging the reading of Voltaire in place of Virgil.

Although the Great Western Railway Bill finally received the Royal Assent on 31 August 1835, among its 251 clauses was a prohibition on building any line or station within 3 miles of Eton College and an obligation to employ staff to prevent pupils accessing the railway. These men were to be paid for by the company but take orders from the college headmaster. The 3-mile restriction, in practice, prevented the infant railway from building a station at Slough but, by now adept at overcoming obstacles, the GWR made arrangements for passengers to join and leave trains there without actually building a station. The college authorities sought an injunction to prevent this and took their case all the way up to the House of Lords but they lost in the end and Slough got its station in 1840.

Oxford and Cambridge universities also feared that the new railways might put temptations in the way of their scholars and both secured protective powers in the relevant enabling acts. Among other things, these gave university officials free access to the local station and the right to prevent anyone who had not graduated from travelling on any of its trains.

THE FIRST GRAND OPENING

This was the title given by the Eastern Counties Railway (ECR) to the events of Tuesday 18 June 1839 when the directors and their guests were conveyed over the first section of the company's new railway line. Planned to link London and Norwich, and subsequently petering out at Colchester, on opening day it managed only to go as far as Romford. On this important Tuesday the ECR was already showing signs of the lackadaisical attitude which was to become all too apparent in its later management and operation.

Four days after the opening ceremony an account of the event appeared in the *Railway Times* which struck a blow for womenfolk by remarking that the decision to exclude them from the event was 'by no means worthy of imitation'. Another feature which vexed the reporter was the shortage of champagne, which took a long time to appear and was then limited to 'a solitary bottle', whereas supplies on the earlier Croydon and Southampton line openings had been significantly more lavish.

The reporter's complaint was fuelled by the fact that the day was a hot one and the two official trains had started late from the ECR's temporary terminus in Mile End due to the non-arrival of an important guest, the Persian Ambassador. They stopped again at Stratford to wait for him and eventually decided to send an engine and coach back to fetch him but no sooner had it left than he turned up, having come by road when he found that the train had left without him.

When the cavalcade eventually arrived at Romford there was further delay in getting the official banquet started and the *Railway Times* reporter sweltered, champagne-less, in the hot sun. His mood was hardly improved by listening to a speech he could not understand delivered by the ECR chairman, Henry Bosanquet, as well as the added confusion that followed. Being told that the return trains were ready, the chairman asked one speaker to cut his speech short. Obligingly this he did only to be replaced by a local MP going through a lengthy round of fulsome praise for the directors. At last, believing that he could get up and start the move back to the waiting trains, the chairman had to be reminded that no one had recognised the contribution of the railway's chief engineer, John Braithwaite. This meant more speech-making before the return journey could eventually be undertaken to the relief of all the participants who, by now, had had their fill of speeches, cannon fire and the interminable playing of the band

PROTECTIVE CLAUSES

Generally, people are wary of something totally new and this is exactly what the early railways were to British people in the first half of the nineteenth century. One result was that powerful interests were able to secure protection for their concerns either by discussion with the promoters of a new railway or by their opposition to the bill that had to be laid before Parliament. Most new railways had to accept a number of general clauses protecting things like water supplies, Post Office services and the like but several had to make specific provision for more curious situations.

The Aberdeen Railway had to shade any lights it installed near the coast so that they would not mislead shipping while the Dundee & Arbroath Railway had to ensure that smoke did not interfere with the communications between the Bell Rock Lighthouse and shore establishments. Eton College bathers using their regular bathing place at Cuckoo Weir were 'not to be disturbed' by the Great Western Railway and the Kent Coast Railway had similarly to respect the privacy of bathers on Ramsgate Sands.

The provision of screens to make sure trains did not frighten horses used on the adjacent roads was a fairly common provision but a more unusual one was contained in a Liverpool & Manchester Railway Act of 1842, which required the company to erect a fence or wall to shield New Bailey Prison. Later on, Great Western had to underpin the walls of Shrewsbury Gaol to keep its occupants secure and the company had also to be prepared to submit to vibration tests conducted by the Astronomer Royal.

Noise and vibration featured in other protective clauses elsewhere. Trains were not allowed to whistle within 100 yards of the Lord Warden Hotel at Dover or near St Thomas's Hospital. The Great Western was prohibited from shunting during the times of Sunday services in Bristol Cathedral. The North Eastern Railway (NER) had an even more onerous obligation in two instances, being required to lay its rails 'on India rubber or similar substance' in order to deaden train sounds. In one of these cases the protection was for the cemetery at Leeds Parish Church, presumably out of concern for mourners rather than the long-term occupants!

One of the most difficult situations for railway promoters arose if their route passed near or through a country house estate. Many were the provisions imposed to protect the view for which the site had been chosen or had been created by the landowner at great expense. The Furness Railway, for example, was to build its Arnside–Hincaster Junction line for 20 chains on arches or pillars to protect the view from George Edward Wilson's Dallam Tower. Lord Harborough wanted the Midland Railway's Saxby line carried in a tunnel beneath his Cuckoo Plantation and the Halifax & Ovenden line was to be arched over its entire length through the Ackroyd estate. Lines at York and Windsor had to be harmonised with their surroundings, another line had to 'present a neat and substantial appearance' and the Llanelly Railway could not build its line at all until agreement had been reached with the bishop of St David's on whether it was to pass in front of or behind his Abergwili Palace.

THE MP FOR LINCOLN

Colonel Charles de Laet Waldo Sibthorp, elected the Member of Parliament for Lincoln in 1826, spent a lifetime perfecting the art of the negative objection. He railed against free trade, water closets, taxes and the 'lamentable influx ... of foreigners talking gibberish'. Sibthorp certainly did not approve of railways, which he said were 'run by public frauds and private robbers whose nefarious schemes will collapse.' Even Gladstone's Cheap Trains Act gave him no pleasure. What he wanted was an 'Act for the Annihilation of Railways'.

Not that railways were alone in attracting Sibthorp's strictures. 'Reform,' he said, was 'a thing which I detest as I detest the devil' and Sir Joseph Paxton's Crystal Palace was 'a transparent humbug and bauble'. It took all of Paxton's guile to save the building from Sibthorp's enmity when it had to be dismantled and removed from Hyde Park at the end of the Great Exhibition.

THE SODDEN FIELD

Achieving rail access to Weymouth and transit through Dorset and East Devon to Exeter was a source of bitter rivalry. The London & South Western Railway (L&SWR) wanted to expand westwards and the Great Western wanted to protect its access to Exeter via Bristol and Taunton. Several smaller companies got caught up in the struggle between these two emerging railway giants including the Salisbury & Yeovil Railway, which had to endure a significant number of setbacks in getting its scheme off the ground (or on the ground to be more precise). A leading shareholder and unofficial company historian described the ceremony of cutting the first sod in 1856 in the following philosophical terms:

> In keeping with the bitter elements of turbulent hostility which the Salisbury & Yeovil Railway undertaking had had to encounter, but at this time seemed to have subdued, were the deluge of rain, and the bitter blasts of wind, which it flung into the faces of the people who were going to the town of Gillingham to be present at the ceremony of 'turning the first sod.' It seemed as though, human adversaries having done their worst to impede, Nature had now taken up the work of baffling and obstructing. That fickle Jade, the Weather, whirled sheets of water on our heads, blew garments into ribbons, and cast our speeches back in our teeth. And when the sodden field was left, and the party sought some protection from the bitter elements under a large marquee, Pluvius made his unwelcome way through the canvas, and, crowning insult of all! mingled our wine with water. These would have been trials enough to break the hearts of many men, but the promoters of the undertaking had had too much of the buffeting of adversity to be depressed now; and they simply raised the diluted champagne to their mouths with the quiet remark, 'We have had so much cold water thrown upon us before that a bucket or two extra can make no difference now.'

THE MIDDY

This was the affectionate name given to the Mid-Suffolk Light Railway which began life at the end of the nineteenth century as an ambitious scheme to fill a gap in the railway network in mid-Suffolk. It was to run from Westerfield on the Ipswich–Felixstowe line of the Great Eastern Railway (GER) to Halesworth on the East Suffolk line and connect there with the Southwold Railway's route to the coast. Another Mid-Suffolk line was to run west from this main line to Haughley as part of a vague dream of providing a through route from Southwold to the Midlands. Apart from a short length south from the planned junction of the two original routes at Kenton, this secondary line was the only one to be built and even then it petered out in a field just over 19 miles east of its starting point at Haughley.

After all sorts of difficulties the line from Haughley to Laxfield was finally opened for goods traffic in 1904 but there was only one train each way daily and the stations were only open when it called. As the official notice stated, 'Your attention is particularly directed to the fact that at the present the stations will only be open at specified times … and that Traffic can only be dealt with at those times.' Small wonder that revenue was insufficient to service the huge capital cost of £205,887. Along with the bankruptcy of the Middy's chairman leaving £20,000 of worthless promissory notes and the alienation of its Great Eastern neighbour who suspected him of links with the predatory Midland Railway (MR), the company's prospects looked very poor indeed. The appointment of a Receiver in 1907 and the introduction of a passenger service in the following year somehow kept it afloat until eventual incorporation in the London & North Eastern Railway (LNER) under the 1923 Grouping.

Four years after rail nationalisation in 1948, the Mid-Suffolk line was scheduled for closure prompting a trip by Geoff Body on 12 July 1952, just two weeks before the end. First the Middy had to be reached from Sandy, the junction between the East Coast Main Line and the Bletchley–Cambridge route. His notes record:

> Left Sandy on the 7.55 a.m. for Cambridge. This line is now part of the Eastern Region between Willington and Cambridge but although the latter's locomotives undertake a few of the workings, London Midland stock is used and Bletchley locomotives predominate. My train was worked by LMS class 2 2-6-2 No.42062 and consisted of five coaches. The leading one was an open coach divided into smoking and non-smoking saloons by a doored vestibule in the centre. At one end was a driver's compartment, at the other a luggage section. Numbered 3411, it had fold-down steps which locked the train brake when in the lowered position and was used at the halts which had no platforms. The reversible knifeboard seats were brass bound and stamped LNWR.

Cambridge shed had a mixed bag of loco types, which, along with the station's exceptional main platform, made the location interesting.

The mid-Suffolk Light Railway neither needed nor could afford elaborate stations as the image of Laxfield post-closure makes clear. Hopes of continuing the line eastwards also expired here.

Eight years earlier, in 1952, Class J15 locomotive No. 65388 takes a breather in a distinctly frail-looking shed.

D16/3-Class 4-4-0 No.62536 headed my semi-fast train on through Newmarket and Bury St Edmunds to Stowmarket. From there it was just a short trip on a Norwich stopping service to get to Haugley and cross to the bay platform to join the 11.15 a.m. for Laxfield. This time the locomotive was one of the ubiquitous J15 Class 0-6-0s, No.65467 with just a two-coach train. Branded 'Haughley & Laxfield No.2', and with only part partitions for the compartments in one of the vehicles, it formed the leading section of a mixed train completed by four wagons and a brake van.

The line rose, fell and curved along a switchback course which had not seen the Chapman weedkiller train for quite a long while. Some level crossings were protected only by cattle grid, at others the guard had to open and close the gates or persuade one of the watching local lads to do it for him. At some level crossings the gates were lying broken beside the track. Despite all this we successfully reached the first two stations and then whistled our way into Aspall and Thorndon, where the passenger coaches were part of the shunting operation to leave two wagons in the siding. More gates and whistling through very rural countryside and No.65476 arrived at Kenton, still a passing point but with little trace of the abandoned route towards Westerfield.

Here some passengers took their near-final chance of being photographed beside the train; no one would have dreamed of going on without them whatever the timetable might say!

More informality at Worlingworth, where the level crossing was protected only by a rope with a red flag hanging from it. An obliging motorist moved the homemade warning device for the departing train and then asked what he should do with it. 'You can have it,' was the reply! On we went, a replacement guard taking over at Stradbroke after a cheerful exchange of greetings in the musical Suffolk dialect, finally reaching our destination at Laxfield.

Like the rest of the Middy's station buildings, those at Laxfield owed much to the inventor of corrugated iron. All spoke of shortages of money with one, little bigger in size than the average garden shed, grandly signed 'Superintendent's Office'. The unusual typeface used for the station name boards was impressive but the engine shed seemed only to remain upright because its wooden sides were held in place by the roof; the roofless portion each side of the inspection pit leaning alarmingly outwards. Another J15, No.65447 stood inside the shed looking slightly bewildered and forgotten.

While 65476 was taking water from a homemade contraption comprising an elevated tank supplied with the help of a small petrol engine, I followed the track eastwards. Soon the standard GER chairs changed to original Mid-Suffolk Light Railway fastenings spiking the rail direct to the sleepers and then even that petered out in a formidable group of thorn bushes, somehow epitomising the failed hopes of the original enterprise. A footplate trip back to Haughley, with 65467 running tender first, rounded off a day of great nostalgia, full of reminders of the wisdom of the closure but also of sadness that nearly a century and a half of community service was coming to an end.

DEODAND

The requirement for anything causing a fatal accident to be considered *deodandum* and forfeited to God had its origins in Saxon times and remained law until its final abolition in 1846. In later medieval years the deodand object might be given to charity or sold to pay for prayers for the victim but, more usually, went to the lord of the manor or the Crown, ostensibly to be put to worthy use. The duty of assessing the value of an item declared deodand fell upon the coroner's jury.

The early railways were not immune to the deodand law, as the Grand Junction Railway found out after a collision between its engines *Basilisk* and *Merlin* caused the death of a driver in 1838 and both were declared forfeit. Three years later the Great Western also made its acquaintance with the ancient deodand practice.

The 2-4-0 engine *Hecla* had only arrived new from its Leeds builders, Fenton, Murray & Jackson, in April 1841 but in the early hours of Christmas

Eve it was involved in a terrible smash in Sonning Cutting near Reading. Brunel had originally intended taking the line through Sonning Hill by means of a tunnel and, with hindsight, this might have avoided the loss of life and the casualties which occurred in the *Hecla* crash.

The Christmas period had been marked by heavy rain, which had caused a large amount of earth to fall from the sides of Sonning Cutting and bury the track completely. Nothing was known of this at Paddington at 4.30 a.m. as *Hecla* left with its train for Bristol. It was one of the two very slow trains of the day on which third-class passengers were carried, and provided only with bare wooden seats and low sides for the nine-hour journey. Behind these open carriages followed a parcels truck and seventeen goods wagons.

With no telegraph to warn of the landslip, *Hecla* smashed into the deep earth in the cutting and the goods wagons in turn smashed into the engine tender causing eight deaths and a variety of nasty injuries. At the inquest, a deodand of £1,000 was decreed as the value of the offending locomotive and carriages, although the GWR did appeal and got this sum greatly reduced.

A DOCKS LINE

When Bristol began to lose trade to other ports because its wharves lay some distance inland along the tidal River Avon its Society of Merchant Venturers promoted the plan for a 'Floating Harbour'. When completed, this provided a new course for the tidal waters of the river and enclosed the wharves by means of locks and dams thus enabling vessels to load and discharge at any time. To serve the activity around the Floating Harbour, a complex of freight lines was subsequently built from Temple Meads to various points on the docks.

Ashton Swing Bridge over the New Cut at the seaward end of Bristol's Floating Harbour complex is no longer needed to give railway access to the docks, but still reveals the sturdy framework of its upper deck.

The original Bristol Harbour Railway was a venture by the GWR, the Bristol & Exeter Railway and Bristol Corporation and probed ¾mile into the dock area, a ½-mile extension being added later. To do so it had to leave Temple Meads at an elevated level between the original station and the goods yard and then use three iron bridges to cross the neighbouring streets. After this section the line sloped down to pass through a graveyard and a tunnel of 282 yards beneath the beautiful church of St Mary Redcliff. To reach Princes Wharf it had to cross the entrance to Bathurst Basin via a steam-operated bascule lifting bridge. Work began in 1868 on the first section and was completed in 1872, the extension opening four years later.

In 1897 an Act was obtained to continue the harbour route with one arm to join the Portishead Branch and another to cross the harbour entrance and come back along the other side of the dock to a huge goods depot at Canon's Marsh. In order to reach the Portishead Branch at Ashton Junction the Bristol Harbour Extension Railway had to cross the New Cut and this involved constructing a double-deck, hydraulically operated swing bridge which cost £70,389. The bridge had an integral signal cabin controlling the double line of railway over the lower deck with road traffic passing to and fro on the upper level.

THE PERFECT LOCOMOTIVE

This was the claim made by one Eli Gilderfluke for a locomotive he 'invented' at the height of the steam era. Clearly intended for American railroads it was of such generous proportions that there was sufficient height for three tiers of driving wheels and space for a number of 'blind' wheels whose sole function was to balance the working motion so that even at speeds of over 100mph the engine would make no more noise than 'a tom cat crossing a wooden bridge'. The three sets of cylinders positioned all over the front end of the locomotive acted on a 'trunk-cross-steeple-tandem-compound' system claimed to achieve a steam use economy of 87.8 per cent.

Mr Gilderfluke's highly imaginary design incorporated what he was pleased to call a 'carbo-wallop', which appears to have been a 'vastly improved' feed pipe to convey exhaust fumes back to the firebox for reuse. Inside the carbo-wallop was a 'deflectorbus' to enable the driver to divert these fumes to the chimney should he so wish. Other special features of the perfect locomotive included a triple X-ray electric searchlight of 9,340 candle power to enable the driver to see ahead, round curves and through tunnels! To discourage wandering people or animals the inventor also incorporated an 'anti-sleep-on-the-track' device.

Altogether this incredible machine was to have over a hundred special features. On paper it was intriguing to look at but would have given locomotive builders, works managers and running shed foremen the worst possible nightmares.

THE ELECTRIC TELEGRAPH

From simple beginnings in 1837, the electric telegraph system was to grow in importance until most railway stations could send and receive domestic messages, with the larger stations also offering this facility to the travelling public. The system was based on using an electric impulse to move a needle pointer left or right to represent the long and short combinations of the Morse code and transmit words in this form along wired circuits which linked groups of stations and which could be connected to extend the coverage of the local system.

The transmitting instrument which eventually evolved, displayed a vertical needle moving through a restricted arc and using a controlling handle located below to achieve this. Moving the needle to the left represented a short 'dot' in Morse and to the right a longer 'dash'. Together these permitted the sending of words and numbers, the whole process being governed by a quite complex set of procedures.

Each station had a call sign and users of the telegraph quickly picked up the rhythm of their own call sign code which was sent when they were required to take an incoming message. Norwich Thorpe, for example, was 'TP'. Holding the needle over so that it could not move was the normal response to indicate readiness to receive an incoming communication. If this was acknowledged by sending 'G' it meant that the receiver was able to receive and translate the coded message into words as fast as it could be sent. To be a 'G Reader' was something of an achievement. Lesser users had to send 'T' after each word to show it had been understood or 'E' if it had not. All messages had a coded prefix to indicate their type and importance, with 'DM', a Danger Message, taking absolute priority. 'SN' meant 'message finished' and 'RD' 'message correctly understood'. There was a book of instructions to cover telegraph regulations and protocol and another containing dozens of code words designed to represent regularly used phrases.

The two pioneers of the electric telegraph system were William Fothergill Cooke and Professor Charles Wheatstone who, in 1837, gave a demonstration of their system on the London & North Western Railway (LNWR) line between Euston and Camden. Four years later the GWR installed this five needle version between Paddington and West Drayton, later changing to a two needle instrument, extending its coverage to Slough and charging the public a shilling to view it in action at the Telegraph Office at Paddington or the Telegraph Cottage at Slough. Advertisements publicising this facility 'respectfully informed the public' that the 'Galvanic and Magneto Electric Telegraph of the Great Western Railway' could transmit 'upwards of 50 signals' to 'a distance of 280,000 miles in ONE MINUTE'.

The value of the electric telegraph for regulating train movements is obvious, but in 1844 it was used to inform London of the birth of Queen Victoria's second son Prince Alfred and also to warn Slough of various pickpockets travelling down from London to events at Eton. Then came an even more dramatic incident in which the electric telegraph proved itself to be much more than a useful novelty.

On New Year's Day 1845, John Tawell set off for Slough with the premeditated intention of committing murder there. His chosen victim was a young woman who had somehow upset him and who was then living in a cottage in the Salt Hill area of the town. His chosen instrument was poison, and in his pocket he carried a quantity of deadly potassium cyanide. Tawell somehow persuaded the unsuspecting target of his murderous intentions to take a glass of stout with him and then managed to slip some of the deadly poison into her drink when she was not looking.

Cyanide acts quickly and the terrified young woman, realising what had happened, let out a piercing scream which caused Tawell to dash from the scene, hurry to the station and board a train back to London. A report of the events that followed records:

> He succeeded in getting into the train and was speedily whirled to London, no doubt flattering himself that he was secure from capture and had evaded the hands of justice. But while on his journey the deed was discovered, and the police of Slough at once thought of trying and testing the powers of the telegraph. Having gained the requisite particulars they sent a message to the police in London, which arrived in time to admit of their being at Paddington Station on the arrival of the train in which the murderer was travelling; and

> from the description given of the man, and the compartment in which he was riding, he was easily detected.

The police did not like to act without a warrant or the proper authority to arrest, and did not consider a telegram sufficient upon which to act. They therefore watched and followed him for two or three more days, and observing his restless conduct, and gestures of apparent distress of mind, and having been, no doubt, by this time armed with the proper authority to arrest him, he was taken, tried, found guilty and deservedly hanged.

It is also recorded that the murderer was dressed 'in the garb of a Quaker' and wearing one of the distinctive long brown greatcoats favoured by Quakers. Unhappily the telegraph instrument at Slough was not able to transmit the letter Q but the telegraph operator still managed to pass on the fullest information by describing Tawell as dressed as a 'Kwaker'. On arrival at Paddington he had taken a bus to the bank and visited several coffee houses, presumably to try to create an alibi, but the plain clothes railway police sergeant watching him was not to be deceived and kept a close eye on his target all the way back to his home in Cannon Street. After conferring with an officer who had come down from Slough, the murderer was duly arrested at his lodgings and began the long road to the gallows.

GENUINE MISTAKES

All systems of communication have their foibles and the railway telegraph was no exception, especially in the early days. Getting just one letter wrong could alter the whole meaning of a message. Turning numbers into letters and using a limited number of dot and dash combinations in Morse just made confusion more likely.

So it was when a stationmaster telegraphed local stations enquiring about a black box that had gone missing and got reports back that no one had seen the black 'boy'. A passenger who had a telegraph message sent to his home asking for his gig to meet him on arrival at his destination station was greatly disconcerted to have his groom hand him a hat box with his 'wig' inside. The mistake probably resulted from the fact that the Morse for a 'G' is two dashes and a dot and for 'W' is a dot and two dashes.

Especially embarrassing was the message sent by a gentleman trying to arrange an event for his circle, who asked the telegraph office to contact a hotel requesting a booking for the twenty-eight gentlemen and five ladies of the party. Remembering that numbers got sent as words it is easy to understand the way the message got transposed into a request to 'secure rooms for two tight gentlemen and five ladies'.

A BRIDGE TOO NEAR

Norfolk has a more than usual share of waterways, something the original railway builders had to contend with and later engineers have had to learn to live with. In the case of the navigable rivers, swing bridges had to be installed which could carry trains when in the closed position but be opened to meet the requirements of shipping. One important swing bridge still in use lies just beyond Reedham and carries the Norwich–Lowestoft line over the River Yare.

After the years of the Norfolk wherries, coastal steamers continued to operate on the Yare between Yarmouth and Norwich and were still frequent visitors even as more and more holiday boats appeared on the rivers and Broads. The Yare was used by the latter not only to reach Yarmouth but also to take the Haddiscoe Cut to access the River Waveney and beyond so that an unhealthy mix of yachts, motor cruisers and 500-ton coasters was not unusual. In theory, power should yield to sail but things are not that simple when sheer size, winds and tides all play their part.

This mix brought the British Rail (BR) engineers out to Reedham Swing Bridge after the 534-ton *Ellen M* hit the structure with her stern on Friday 10 September 1965. The vessel was on her way downstream to Yarmouth and the bridge had been opened for her in the normal way but, just as she reached it, a yacht named the *Four of Hearts* drifted across her path out of control. The helmsman on the Metcalf Motor Coasters steamer did the only thing he could and swung his helm over, managing to avoid running the yacht down but unable to avoid scraping the bridge girders.

One of the four young Londoners on the *Four of Hearts* later described the events:

> We were going up to Norwich from Berney Arms. Just short of the bridge we lowered our sails, intending to moor, but we missed the mooring. We saw this great thing coming toward us. We tried to start the auxiliary engine but the battery was flat. We were drifting helplessly through the bridge with the boat bearing down on us. We were scared to death. The ship swung to starboard to avoid us and slid against the bridge. It was right on top of us and we were fending it off with our feet with the hot water from its engine pouring down on us.

After this brush with disaster both vessels were able to moor, with the coaster then resuming her journey downstream. BR engineers were summoned and soon had the bridge working again but there was trouble once more in the following week possibly as a result of the *Ethel M* incident.

The relief signalman on duty had opened the swing bridge to allow through the coaster *Trilby*, which was returning empty from Norwich, and the *David M*, which was taking coal up to the city. It closed again without trouble but then the wedges onto which it is lowered did not slide back into the locked position. Trains could not be allowed to pass over the bridge until the fault had been rectified so they were halted at the preceding signal boxes, informed of the situation and then allowed to approach the bridge under caution. Passengers were able to alight and walk over the structure to join a train on the other side and continue their journey.

STRANDED

A prelude to the construction of the Severn Tunnel was a lengthy period of survey work to determine the best course for its passage under the widening waters at the mouth of the Severn River. The choice of route was limited by the existing lines from Bristol out to the New Passage Ferry and, on the west bank, by the need to connect with the 1850 route of the South Wales Railway from Gloucester to Cardiff and Swansea. In between lay a varied geological profile beneath the river, dominated by a wide bed of red sandstone, exposed at low tides and known as the English Stones. Then a narrow channel, known as The Shoots, which had been worn deep by the erosion of large pieces of gravel, rushed through it by the great tides and storms of the Severn Estuary.

A team of two surveyors and their staff holder were assigned to trudge around the slippery surface of the English Stones and take the essential measurements. They were taken to the site by two local boatmen, who were under orders to pick them up again at a specific time determined by the next rising tide. The boat was supposed to wait nearby but, for some reason, the boatmen stood off and then found that, row as they might, the strong stream flowing through The Shoots was too powerful for them to make the agreed rendezvous. Instead they were swept helplessly away and finished, exhausted but thankful to be alive, at Avonmouth.

Meanwhile, their day's levelling work finished, the engineers trudged over to the spot appointed for their pickup only to find no boat in sight. Around their boots the water was already lapping and they knew that its rise would soon be rapid. They all began shouting for help from the shore, combining their efforts to try for the greatest effect but attracting no attention because of the natural noise of the moving waters. Young Dan, the staff holder, was by now shouting in panic, and making himself hoarse, such was his fear of the clearly rising tide.

There was a happy ending to this incident thanks to someone on shore realising that the surveyors' boat had not returned and sending another one for them, but it had been a very near thing.

In this same area, back in the Civil War, a troop of Cromwell's soldiers had been less fortunate. From the triumph of taking Raglan Castle, some sixty or so were sent to pursue the fleeing Royalists across the Severn and had commandeered a ferry to take them over to the English shore. Sadly for the eager pursuers, the sympathies of the ferrymen lay with their enemy and the soldiers, not knowing the treacherous nature of the lower Severn, were deceived into landing on the Black Rock from which they were told they could easily gain the shore. Instead they were all drowned by the merciless rise of the Severn tide.

TROUBLED TRIAL

Once the worst years of inter-railway competition were over, the individual railway companies began to cooperate more. Locomotive departments not only exchanged ideas and technical data but, from time to time, joined in practical working tests. In 1921 one of these involved testing the hauling powers of a GWR 2-8-0 goods engine design on a particularly difficult section of the North British system. This was on the NBR main line from Perth to Edinburgh, which, between the Bridge of Earn and Glenfarg stations, presented the challenge of a winding climb of 6¾ miles at a gradient of 1 in 75 through two tunnels to the top of Balmanno Hill.

The chosen GWR locomotive was duly worked up from Swindon and for the first test was given the task of lifting a 600-ton train of mineral wagons up the steep gradient in a time of thirty-three minutes. A worse day could not have been chosen for snow was falling heavily when the time came to start and a strong wind was blowing the snow into drifts. Driver Lovesey and Fireman Cookram needed all their skills to get the heavy train moving and then achieve the required speed but the Great Western engine breasted the summit exactly on schedule to the warm approval of the observers.

Not content with this excellent performance a decision was taken to add five extra wagons and run the test again. The load was now 683 tons and the weather was even worse. The 2-8-0 slipped badly on starting and then encountered patches of frozen snow on the rails as soon as it hit the rising gradient. Sanding worked at first but the heavy snow soon blocked the sand pipes and halfway up the merciless climb the test train ground to a halt. The powerful superheated Swindon engine had done her best but had to concede defeat to the Scottish weather.

EARLY COACHES

As on most railways, the early passenger coaches on the London & South Western Railway clearly showed their stagecoach origins. Like their road coach predecessors they had four wheels and little or no springing and the first-class vehicles were exactly like three stagecoach bodies joined together.

The similarity even extended to a variety in the external appearance; at one period the livery might be a bilious greeny-yellow, a concoction of chocolate and salmon or just plain varnished wood. One major difference was that the L&SW guards' vans had their ends painted a strong scarlet colour, presumably to discourage any following train from running into them.

The class of travel was not only reflected in the provision for comfort – hardly any in the third-class vehicles – but also in the fitments. The first-class compartments had an oil lamp in each, second-class travellers enjoyed one lamp between two while in the third class just one lamp was suspended through a hole in the roof to shed its feeble illumination over the half-height partitions of all four sections.

THE FIRST STEAM WHISTLE

The origin of the locomotive steam whistle was described in a Yorkshire newspaper report which appeared many years ago:

> One morning in 1833 the engine Samson ran into a horse and cart crossing the line at Thornton, on the Leicester & Swannington Railway. The engineman, having but the usual horn, could not attract attention. This resulted in having an appliance made by a local musical instrument maker. It was put on in a few days, and tried in the presence of the Board of Directors, who ordered other trumpets to be made for all the engines the company possessed.

The steam trumpet described was apparently 1ft 6in in height and 6in in diameter at the top.

SLIPS AND FUMES

Most railway construction involved unforeseen problems, even later schemes like the 1905–07 work on the Birmingham & North Warwickshire Railway (B&NWR), a GWR project to give the company a shorter route between Bristol and Birmingham.

The Great Western's need stemmed from the company being out-manoeuvred by the Midland Railway's acquisition of the infant Bristol & Gloucester concern back in 1845. After that, getting from Bristol to Birmingham over its own metals involved the GWR in a roundabout journey of 133 miles via the Severn Tunnel, Hereford and Worcester. Now the new railway it had promoted would alter all that. Along with another new link between Honeybourne and Cheltenham, the 17¾-mile B&NWR line between Bearley and Tyseley would cut 38 miles off the journey.

There were cuttings to be made over half the length of the B&NWR with 1.5 billion cubic yards of soil to be excavated altogether. As usual this would be used to form embankments. A steam navvy started on the first cutting at the north end of the route in February 1906 and soon had the spoil built into a 14ft embankment, which then consolidated nicely as men and machines continually passed over it. However, when the next layer was placed on top, the loam and water still held in the lower section acted as a lubricant and the whole mass quickly spread outwards and lost it shape. The contractors now had the extra task of stabilising the lower slopes with slag and using loads of ash to lighten the top.

Work on the 175-yd Wood End tunnel also started in 1906. There was only half as much excavation to be done at the south end as at the north so the clever plan was to start there, then bore the tunnel and bring the spoil from the northern cutting out through it, a much easier task than carting it over the top of the hill. Everything went smoothly for five months until the blasting fumes overcame the fans in the tunnel and an unplanned ventilation shaft had to be dug to the surface.

The delay was minimised thanks to the Herculean efforts of one man who dug his way through 40ft of marl to the surface in one long night shift. His satisfaction must have been even higher a week later when the roof of a heading collapsed and the tunnelling crew were able to escape by scrambling up the new shaft to safety.

This line also involved the challenge of a unique elliptical skew bridge but all the work was successfully completed after twenty-six months and the new piece of railway carried its first train on 9 December 1907.

RAIL-AIR 1883-STYLE

The Great Western's pioneering work in setting up air services to fill gaps in the rail network is well known but the company apparently had a modest involvement with air transport much earlier than thought. A brief but intriguing report in the *Slough and Windsor Express* of 1 September 1883 describes an occasion when the GWR participated in an experiment with what was described as a 'steam sailing machine'.

At this time the Staines & West Drayton Railway (S&WDR) was under construction starting with the section between West Drayton and Colnbrook. The branch was a Great Western project with the company working the line from its eventual opening in 1885 and absorbing the S&WDR in 1900. It seems that an inventor from Margate approached the railway authorities with a request for assistance with his project to launch a steam flying machine and, rather surprisingly, got a positive response.

On a section of the newly laid track the inventor's brainchild was duly mounted on a flat truck. The contraption consisted of a frame of light wood fitted with two large wheels at the front and two smaller ones behind. It was to be fed with steam from the propelling locomotive and this would work 'a nine-bladed screw' with 'gearing manipulated from another truck'. To further quote the report:

> The inventor's idea is to propel the machine by steam on land until it attains a speed of 30 or 35 miles, a velocity which Mr Lynfield (the inventor) calculates will be sufficient to lift the machine into the air, when it will be navigated by means of the sails with which it is fitted.

The optimistic Mr Lynfield had expressed a conviction that it was 'possible to fly in the air at the height of a mile from the ground.' But he had to be satisfied with something a little less ambitious, described modestly by the reporter thus: 'The operator succeeded in getting the machine lifted from the truck into the air and expressed himself fully satisfied with the result of the trial.' The GWR's locomotive superintendent had watched the proceedings with interest but his reactions were not recorded!

OFFICIALLY CURIOUS

After obtaining its enabling act in 1846, the Manchester, Buxton, Matlock & Midlands Junction Railway subsequently went back to Parliament seeking authority for a succession of deviations to the original route. All the various changes were recorded by the examining Parliamentary Committee, in 1848, when the overall picture was so peculiar that the committee, with heavy sarcasm, observed of the latest plan:

> It cannot be denied that a line which runs through the lawn before a nobleman's house, dashes through an unnecessary tunnel of two miles, at an unnecessary expense of £250,000, and places the principal town from which it derives its name upon a branch with a gradient of 1 in 20, is a 'railway curiosity'.

THE ATMOSPHERIC CAPER

After seeing it at work in Ireland and receiving Brunel's endorsement, the Board of the South Devon Railway decided to use Samuda Bros' atmospheric system for working the new railway, which they had been authorised to build between Exeter and Plymouth in 1844. In essence the system involved trains being connected to a piston running in a continuous pipe between the rails and being drawn along as a result of stationary engines creating a vacuum in front of the piston.

Advocates of the atmospheric system confidently predicted big cost savings as a result of raising steam at pumping stations along the route rather than having to purchase, man and maintain individual locomotives. Unfortunately, the reality was very different. For over a year after the opening of the line between Exeter and Teignmouth, locomotives had to be hired to work the trains and even when atmospheric working began on 13 September 1847 a variety of practical problems soon began to surface.

On the Starcross side of the River Exe, the ferry from Exmouth landed its passengers at this berth from which it was just a few yards to the station. The former atmospheric railway pumping house building dominates the background.

One area of difficulty was the system's lack of flexibility. Even reversing the direction of trains involved unbolting the piston carriage from the hauling platform and reversing it. Wear was another problem, with the leather piston plug getting quickly and badly worn when it passed the pipe inlet and outlet valves. The coulter plate, which projected through the slot in the pipe, also suffered greatly as it made its unlubricated passage back and forth along the route

Even more of a problem was leakage from the piston valve slot in the pipe. The sealing arrangement was based on covering the slot with a leather seal fixed on one side and kept airtight on the other by lime soap. This proved unsatisfactory due to a reaction with tannin in the leather and the salt-laden atmosphere of the coastal route. Whale oil was substituted, but this attracted vermin, which seemed only too anxious to feed on the oil-soaked leather even at the risk of being sucked into the pipe.

The loss of vacuum due to leakage made the train services unreliable and meant that the pumping stations frequently had to put both their engines to work, with a resultant increase in the fuel costs. There was also a tendency for the morning start-up to lead to a shower of dead rats and dirty water being exhausted from the tube into the engine houses.

Despite all these tribulations, atmospheric working between Exeter and Teignmouth continued until 30 June 1849 and performed quite well at times. A top speed of 68mph was achieved on one occasion but the unplanned extra costs of fuel and repairs, and the fundamental weaknesses of the whole concept, finally brought a decision to change to conventional operation and the end of a bold experiment.

FEEDING THE HABIT

In the early 1930s, the Great Western Railway fitted miniature chocolate machines in sixteen of its crack expresses, including the *Cornish Riviera* and the *Torbay Express*.

Only 18in long, 4in wide and 2in deep, the new machines were painted to match the general coach décor and provided with a mirror front.

DONCASTER RACES

In the days of the Big Four railways, special arrangements were always made to cater for the crowds travelling to and from any major event. Securing optional travel to rail was an ingrained part of the commercial culture and only in later years were the economics of keeping huge numbers of coaches idle between excursions taken into account. Doncaster Race Week in 1929 is a good example of the extraordinary rail activities on these special occasions.

On St Leger Day Doncaster Central station dealt with eighty-nine train arrivals between 8 a.m. and 1.30 p.m., fifty-two of them 'race specials'. Between them these trains brought in over 34,000 passengers. A similar number would return to their starting points in the evening so space had to be found in the locomotive sidings and other places to keep the empty stock until it was required for the return journey. The locomotives had to take coal and water and be manned as necessary. In all over the week something like 200,000 people would be conveyed by rail to enjoy the racing programme.

In addition to the horse-racing programme itself the Doncaster event was also a prime venue for horse sales. Many of the horses came by rail in seventy special trains for which over 1,000 horse boxes had to be found. Altogether 1,366 horses were dealt with including a considerable number of yearlings for Tattersall's Bloodstock Sales.

The background work to ensure the railway operations went well necessitated bringing in extra staff to help with the vastly increased work, including that of examining tickets when the incoming specials were stopped at ticket platforms north and south of the main station. The engineer had to erect notice boards, provide extra lighting, arrange crossing points and even organise more lavatories. Everything was planned down to the finest detail including the supply of more rock cakes and extra tea urns for the refreshment rooms.

SEASON TICKET PERKS

On an early Irish railway, seeking to stimulate traffic, season ticket holders were granted the privilege of the free use of the Company's cold-sea bathing places on 'weekdays only'. Mind you the same company warned passengers that it 'does not hold itself responsible for any interruption that may take place in the intercourse along the railway.'

THE MADNESS

For seven years after the opening of the Liverpool & Manchester Railway in 1830, the network of railways in the British Isles had grown steadily. In the years 1833–37, Parliament authorised 2,050 miles of new railway, nearly five times the length already opened. Then came a period of industrial and financial depression which so dampened the enthusiasm and money available for railway schemes that only 245 miles of new line were approved in the five years between 1838–43.

In 1844, the business climate began to change and interest in new railway schemes revived, as a result of a general improvement in trade and the money supply, ushering in a period of madness which later came to be known as the 'Railway Mania'. At the same time, some of the existing railways were paying substantial dividends, 10 per cent in the case of the London & Birmingham and Grand Junction companies and no less than 15 per cent by the Stockton & Darlington. Suddenly everyone with a little spare capital wanted to invest in a railway, prompting new schemes to appear almost daily: some with sound prospects, many based on no more than optimism; some quite reckless, whilst others were blatantly fraudulent. The forty-nine schemes receiving their enabling acts in 1844 heralded a wild period in which no less than 1,213 bills were lodged in the next three years, of which 581 were passed, representing 7,592 miles of new railway.

Behind these figures lies a picture of speculation frenzy. Thousands of ordinary people with private means, or who ran small businesses, were persuaded that railway investment was the chance of a lifetime. Their eagerness to share in the boom led to reckless decisions and the offer of their capital without any investigation of the intended recipients. Unscrupulous manipulators dreamed up worthless projects and took the cash so readily offered, shares often changing hands again for a quick profit within minutes of being acquired. Some of these shady promoters just disappeared, others more foolish than fraudulent, pursued their grand railway projects without the ability to see that not only would the promotion and Parliamentary proceedings be extremely expensive but that their route was going to be hopelessly costly, sparsely trafficked or throttled by competition.

Although so many investors suffered from the Railway Mania, a number of trades and professions concerned with the railway promotion process were to benefit enormously. Every potential line had to be surveyed. Qualified surveyors were in short supply and could not only demand enormous fees but live extremely well while doing their job. As one report put it:

> Some of the survey staffs assumed the importance of regimental headquarters; claret, then an expensive wine, was drunk at breakfast; and the best sherry that the landlord could produce was condemned as not good enough for the young swells who crowded around Mr Theodolite, the chief of the survey.

One of these so-called 'young swells' boasted that he could survey 5 miles of open country in a week and earn £30 a mile in the process. Still more surveyors were later needed until these figures could be doubled. Needless to say, many of the surveys done this way proved vulnerable to cross examination at Parliament's committee stages.

The host of competing railways schemes also pushed up demand in the legal profession. A famous Queen's Counsel (QC) got 1,000 guineas for one crucial speech and in sought-after chambers the clerk could make £3,000 a session in fees. Parliamentary agents, too, did exceptionally well and even parish clerks and notice servers derived an income from the embryo railways. Printers were delighted at the increase in orders for the production of prospectuses and the fortunes of newspapers became much improved by the revenue derived from the obligation to publish railway plans.

Something like 170 engineers were involved in drawing up technical plans for the schemes of the Railway Mania years. The most capable of them had many more appointments than they could manage personally and employed a host of assistants to do the donkey work. Robert Stephenson, for example, had responsibility for no less than thirty-seven schemes and Sir John Rennie only three less. So great was the demand that prospectuses began to appear with the name of the engineer shown as 'to be appointed'.

Another highly lucrative area was in the field of support for and opposition to planned railways during Parliamentary investigation. Local landowners were drawn in, partly by their anxiety to protect their lands and homes from intrusion and then, increasingly, because of the strength of their position in relation to the sale of land required for the new lines. Witnesses, too, did extraordinarily well. Again, this was graphically expressed in a report which averred:

> The country was ransacked for witnesses who would consent to go to London to give evidence, live like noblemen at the best hotels, stay away for three weeks from their shops or warehouses, and return with each a hundred guineas in his pocket.

In these few hectic years of the Railway Mania, well over a thousand schemes were floated covering every area of the country. Countless small investors were ruined, many small businesses foundered and many a worthwhile railway project was ruined or delayed because it was denied capital because of the false promises of the fools or knaves seeking only to get rich quick without the bother of sense or scruples.

ROAD/RAIL COORDINATION

One of the objectives underlying the nationalisation of the railways in 1948, and the subsequent developments under the British Transport Commission, was the integration and coordination of the various forms of public transport. Not that this was a new idea, especially after railways gained powers at various periods to operate ships, aircraft, buses and lorries. But one example of a practical link between two transport agencies occurred much earlier than most people think in that stagecoaches on the London to Exeter route were using rail flat wagons for part of their journey very soon after the first sections of the London & Southampton Railway were opened.

In the 1830s, Exeter was well served by mail and stagecoach services to and from London. On the two different routes there were no less than seven daily coaches with others operating two or three times a week. The *Telegraph*, *Quicksilver Mail*, *Subscription*, *Defiance* and *Vivid* were well-known and respected coaches, which linked Devon with the capital in nineteen hours, despite the hazards of weather and the shortcomings of road surfaces.

The world of the coachmasters was thrown into turmoil with the advent of the railway but many adapted to the new technology by altering their routes to serve the advancing railheads, or to provide cross-country feeder services. W.J. Chaplin went one better: he sold his huge coach business and invested in the new railway. At least one other proprietor did a deal to use flat wagons to convey his coaches to a railhead and then complete the journey by road.

Sherman's *Telegraph* used the Nine Elms to Basingstoke section of the line and then continued to Exeter by road, which enabled the coach to cut the overall journey time by two hours. The ordinary first- and second-class fares were paid for the rail journey and the coach driver or guard was permitted to travel free.

The *Subscription* coach was another one to use the new railway. After receiving its enabling act in 1834, the London & Southampton Railway had opened as far as Winchfield by September 1838. A newspaper report from the following February reveals that it was already carrying the Exeter coach by then. Unfortunately the report concerned a 'DREADFUL ACCIDENT' and read thus:

> We regret to state that a melancholy accident, which proved fatal, happened to Mr Charles Hex, guard of the Subscription coach on the Southampton Railway, on Friday, 15th inst., as the train was proceeding from Farnborough to Hartley Row. Hex was informed by the conductor of the train that the luggage on the top of his coach had shifted; and as it is usual for guards of stage coaches to ride inside their own coach while on the train, he came out

as the train was proceeding, in order to secure it; when in the act of returning from doing so, it is supposed his head came in contact with the springing of one of the arches, which hurled him with great violence to the ground. One of the conductors saw him fall, and instantly gave the alarm; but the noise of the engine being so great, the engineer was not able to hear it. As soon as the train had reached its destination, Inspector G.R.Gunnell despatched an engine and carriage, with T.Carter coachman of the Exeter Telegraph and Johns, coachman of the Swiftsure, who found Hex about two miles and a half from the terminus, with his skull much fractured, and senseless, bleeding on the ground; the loss of blood was most profuse. He was then conveyed to Mr John Webb's, Murrell Green, where three medical men were immediately in attendance, and all possible means adopted to restore him, but without effect; he lingered in a senseless state till Saturday two o'clock when he expired.

THE MERCHANDISE CLASSES

In their later years, railways tended to be greatly overregulated. As common carriers they were obliged to carry pretty much every category of merchandise and their freight rates were regulated to a point which severely fettered the industry in its battle to compete with the growing power of road transport. Railway freight rates had to be open for scrutiny and this, together with being forbidden to display 'undue preference' to anyone, meant that railway rates clerks and canvassers had very few cards to play in their negotiations with traders. The freight rating system also changed very little over a century with the pre-nationalisation *Classification of Merchandise* book still displaying its origins in documents like the 1854 *Classification of Goods Merchandise applying to all stations upon the Eastern Counties, Eastern Union, Norfolk, Newmarket, London and Tilbury Railways and also the Royston and Hitchin Branch Railway.*

This 1854 document allocated goods to a mineral class, a special class or one of five other groups which governed the rates that would be charged for their conveyance. The commodities specified within each group reflected their nature, with the coprolites, culm and dross in the mineral class contrasting with the bonnets, clocks and elephant's teeth in the fifth class. The rates for each group would be related to their weight, bulk and value.

The mineral class was basically for commodities like bricks, gravel and sand but dross, loose manure and soot appeared there too. As well as coal, iron and grain, some curious items like ashphaltum, mangelwurzels and guano came in the special class. The list of items in the first class embraced a huge range of products from candles to cannon balls. Along with 'Horn Tips, packed', it catered for such things as *divi divi* and *shumae*, whatever they were. Cotton, cheese and hay were charged at second-class rates, together with bass mats, bedsteads and broomheads, while beeswax and bellows were in another large group, the third, and gig shafts along with turtles in the fourth.

A few items were treated separately including rowing boats, where movement cost about 10*s* for every 50 miles. The charges for furniture in vans had to include the weight of the van, while timber charges were based on a table, which converted length to tonnage for each type of tree. Gunpowder was conveyed from London only, while the official notice made the clear statement, 'Carriages, Gigs and other Vehicles for the conveyance of Passengers not carried by Goods Train on *any* terms.'

DICKENS TO THE RESCUE

Charles Dickens was travelling on the South Eastern Railway on 9 June 1865 when he was involved in what he called 'a terribly destructive accident'. This occurred to a London-bound train as it passed over the viaduct across the River Beult on the approach to Staplehurst station. The locomotive was derailed but fortunately the first coach, in which Dickens was travelling, stayed on the track. To his great credit he did what he could to help those in need of assistance, rescued his travelling companions 'much soiled, but otherwise unhurt', and only then went back for the manuscript of *Our Mutual Friend*.

HAIR RAISING

Queen Victoria's endorsement of train travel seems to have influenced her household and on one occasion saved her hairdresser, Mr Isidore, from the real prospect of losing his job. After a trip up from Windsor to London he missed the return train, which would have got him back in time for Her Majesty's second daily session. With two hours to wait before the next scheduled service, he had no alternative but to ask for a special.

For £18 Mr Isidore got his train and so impressed the GWR management with the urgency of the occasion that they ordered 'extra steam to be put on'. As a result, the 18-mile journey was accomplished in eighteen minutes, which probably increased the hairdresser's state of anxiety but saved the Queen's coiffeur and Mr Isidore his job.

IF YOU CAN'T BEAT 'EM ...

The opening of the Liverpool & Manchester Railway in 1830 affected only a few of the many thousands engaged in the horse-drawn coach and wagon business. A decade later the position was vastly different and a host of long-established trades, from coachbuilder to innkeeper, viewed the increasing railway activity with fear.

Typical of the age, however, public and operators alike were quick to react. As early as 1839, a Cotswold parson records taking the usual London coach but transferring to a train at Maidenhead. There, using a special access road, the stage coach was loaded onto a flat wagon and attached to a Paddington train in which the clergyman was carried first class. Arriving fifty minutes later, two fresh horses hauled the unloaded coach to its usual terminating point. Two years after this, the same clergyman took his own coach by rail from Cirencester to Chippenham, including a change of trains at Swindon.

As the Great Western Railway and its Bristol & Exeter partner advanced into the South West, most of the affected coach operators gave up their through services in favour of providing cross-country connections, but one started using the new train service within a few months of its introduction. In August 1841, the *Exeter Flying Post* carried an advertisement announcing:

THE DEFIANCE TO LONDON
EVERY EVENING AT SIX O'CLOCK
from the
HALF MOON HOTEL, EXETER
via
BRIDGEWATER AND THE GREAT WESTERN
RAILWAY
IN ELEVEN HOURS!!!

Forsaking the Wincanton–Andover route, the revised service saved its patrons five hours.

LET THERE BE LIGHT

In the early years, rail travel at night was limited and very uncomfortable. If there was any lighting at all it was likely to be a candle and passengers were generally expected to provide their own. They could be bought at the station inn or from one of the early bookstalls. Railway provision, when it came, was limited to oil lamps, much given to smoking when the wick got sooted up.

An odd example of the use of oil lamps was on the 12½-mile Isle of Wight Railway. Its only line terminated at Ventnor after emerging from a steep descent through a tunnel into an area hewn from a rocky cliff. When a train from Ryde stopped at Wroxall, the preceding station, porters would open each door and hang an oil lamp on the inside to provide illumination through the tunnel.

Gas lighting was a big step forward, the gas being manufactured from oil shale and moved in a pyramid of cylinders on special wagons to the station storage, from which it was then piped to reservoirs beneath the coaches. More piping then took it to the roof lights, which were fitted with mantles for incandescent burning. Lighting was then operated by using the chain to turn a simple valve in the bypass supply. All very well but a significant fire risk from which many an accident victim suffered awful incineration.

Despite the change to electricity, Tilley lamps still lit some smaller stations and the paraffin platform, hand and signal lamps continued in service until relatively recent times.

BLACK DOG

Nineteenth-century landowners were notably fond of demanding privileges in return for allowing railways access to their property. Their requests ranged from special stations and stops to an expensive tunnel to protect their view. Black Dog Halt in Wiltshire was a typical example. It lay on the Calne branch from Chippenham and was built to meet the wishes of the Marquis of Lansdowne and serve his Bowood House and estate.

Black Dog Halt had two sidings, one of which was used for loading racehorses and the other serving the more domestic purpose of conveying household items between Lord Lansdowne's local residence and the one in London. The halt had a ticket office and a stationmaster, the Marquis insisting that the latter should have no political views or affiliations. In return for obtaining aristocratic approval the man appointed was provided with free housing and fuel while half of his salary also came from Bowood House. Despite its facilities, Black Dog Halt was not in the public timetables and was, effectively, a private station which locals were allowed to use.

These feudal arrangements gradually altered, with the station becoming a public facility in 1904 and the Halt elevated to normal station status in 1953. It was never of major consequence: the stationmaster being replaced with a leading porter, the service operating on a request stop basis and then eventual closure in 1965.

SLEEPING ARRANGEMENTS

The final conversion of the GWR's broad gauge to narrow, or standard, was preceded by two phases, one the addition of a third rail to create dual gauge on some routes and the other the complete changeover on some associated branch lines. An early replacement of broad by narrow gauge track took place on the line from Grange Court Junction to Hereford via Ross-on-Wye – probably to gain experience of the process on a quieter, less busy track.

For two weeks, from Monday 16 August 1869, train services on the branch were scheduled for displacement by horse buses. Two days earlier, three special trains conveying 400 workmen arrived from South Wales, each bringing a week's supply of food with him. The intention was to work continuously until the job was finished and provision for sleeping was made by providing forty broad-gauge wagons, which had been specially cleaned, whitewashed inside and then lined with straw to create bedding for the workmen. Those in charge had the relative luxury of a first-class carriage.

Conversion work on the 22-mile route went so well that it was completed by the Thursday, allowing the first narrow gauge train to run on the following day. The Friday was spent completing the works in station yards and tidying up, allowing the men to be paid and return home the following day. Altogether, quite an achievement.

DUBIOUS FIRSTS

Like many pioneering enterprises, the Swansea & Mumbles Railway did not have it easy. This very early line was authorised in 1804 as an 'L' plate mineral tramroad between Swansea and Oystermouth to carry stone for building work in the former. It opened in April 1806 and eleven months later an enterprising chap named Benjamin French began a passenger service over the route using a horse-drawn wagon. This was quite possibly the world's first regular railway passenger service. Again, quite possibly, a less prestigious distinction is that of being the first passenger-train operation to succumb to road competition.

In 1826 a turnpike trust built a new road to parallel the tramroad. It was well constructed with a good surface and the coaches which began to use it provided a much more comfortable service than Mr French's vehicles.

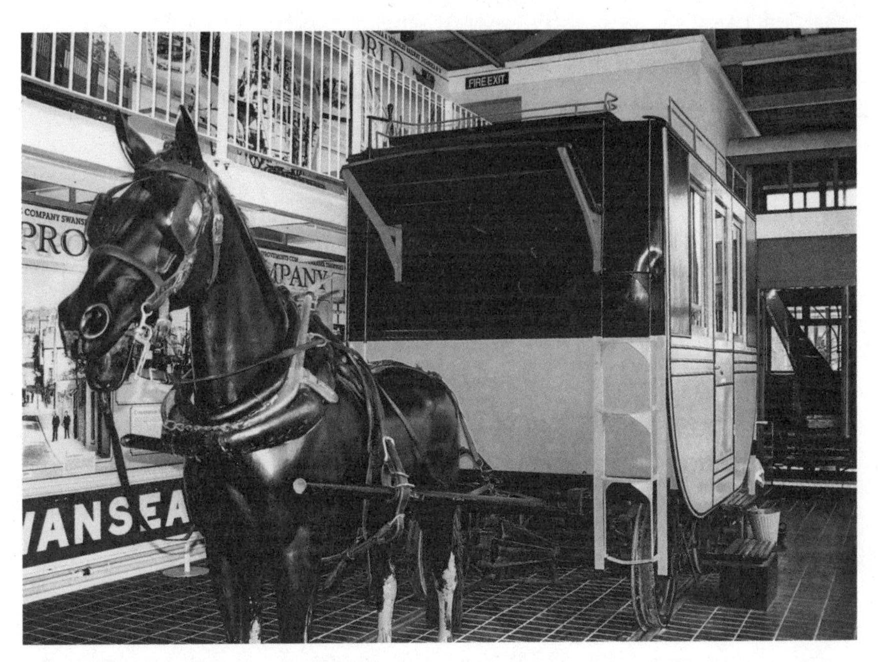

This tramway carriage looks fine in its Swansea museum setting but clearly has no springs and would offer only an uncomfortable ride.

He had improved their design but, running as they did on their very basic tramroad metals, they remained rather slow and prone to jolts and other discomforts from unforgiving seats. Their patrons transferred to the new road rivals and the rail service ended.

Later, the Oystermouth line got its rail passenger service back, with horse traction replaced by a motley collection of locomotives and two-deck tram-style vehicles and with the line also extended to a Mumbles terminus. For a time the engines carried a youth perched on the leading buffer beam, presumably to warn of any obstruction along the line. The enterprise then got more interference when pushed off part of its alignment in 1867 by the new Llanelli Railway extension.

A new era seemed to have dawned for the Swansea & Mumbles Railway when, in 1928–29, it was electrified for normal tram operation, with electric current collected via a pantograph on the top of the vehicle. There were some very busy periods, especially at holiday times. The operator, however, was a bus company and the Swansea & Mumbles notched up another probable first when the end section was turned into a bus lane. The final blow came in 1959 when the line closed and a replacement bus service took over.

GANGS

As in many other walks of life, the railway industry had its gangs. There were permanent way gangs, of course; also the much-feared 'Razor Gang', whose visits presaged the axing of some unfortunate line, job or facility. A notable, but less ubiquitous gang occupied accommodation left over from an old lime kiln on the sea front east of Teignmouth, in South Devon. This was the 'Cliff Gang', whose job it was to cut back vegetation on the red sandstone cliffs that back the coastal rail route from Exeter to Teignmouth and Newton

Abbot. In steam days this cliff vegetation was at risk of being set on fire by locomotive exhaust sparks and, with no easy access to the area for firefighters, prevention was deemed better than cure. Hence the 'Cliff Gang'.

Another unusual gang worked at the small railway wharf at Dunball, on Somerset's River Parrett. Siding-connected with the main Bristol to Exeter line which ran nearby, this modest harbour was, for many years, busy with the loads of coal and other goods brought in by sailing trows and ketches, and later by small steamers. To give the vessels a flat bottom to sit on at low tide, a gang of 'Mudders' was employed to level the wharfside mud, a dirty and thankless task with few perks except their own Mudders' Hut in which to clean up and a separate room in the nearby pub for a restorative of a different kind, usually local cider.

TROUBLE

The modest single-line branch from Kemble to Tetbury closed in 1964. The closure spelled the end of the intriguingly named Trouble House Halt. The tiny wooden platform there took its name from an inn on the nearby road running from Cirencester to join the A46 north of Bath.

Still meeting the needs of travellers and local people, the Trouble House Inn was originally built by a Cirencester carpenter in 1755. Not long afterwards, the troubles that led to the inn's unusual name manifested themselves when several landladies died young and two landlords came to an unnatural end. The inn lost customers to press gangs seeking recruits for the American war and then got burned down in one of the riots against the introduction of agricultural machinery. Apparently some of the spirits made restless by this run of misfortunes began to haunt the place and its old name of Waggon and Horses gave way to the present one. Fortunately, bad luck did not rub off on its companion railway halt.

GLOUCESTER

Due to early railway rivalries and competition, many of the larger towns and cities of Britain had more stations than they needed. Gloucester was one of these, although its two stations were, at least, well situated in relation to the city centre.

The Great Western Railway had planned to acquire the Bristol & Gloucester Railway in order to spearhead its territorial ambition to extend north towards Birmingham. However, it procrastinated in the negotiations and the Midland Railway (MR) nimbly got in first. An early result of this was the Great Western's need to provide a curious 'T station' on its route around Gloucester. Here, it installed a turntable to get carriages into its proper station until the line along the west side of the River Severn to Chepstow was completed. Another result was that Gloucester for many years had two stations, Eastgate on the Midland route and Central on the Great Western.

The railway approach to Eastgate from the south involved a level crossing that was a constant source of irritation and delay to road users, and the combination of local and financial pressures eventually led to the decision to close the MR route between Tuffley Junction and Eastgate. Instead, all services would be concentrated on Central station and routed via the avoiding line. Eastgate was duly closed in 1975 and the main platform at Central extended to allow trains to use both ends. This did mean through cross-country services having to reverse at Central, but seemed a small price to pay for the benefits. Rail users could have fared worse if a contemporary scheme to serve both Gloucester and Cheltenham with a single station midway at Barnwood had come to fruition.

A casualty of this rationalisation at Gloucester was the unusually long footbridge that connected the two stations.

REPLACEMENT BUS SERVICE

The replacement of train services by buses tends to be thought of as a feature of the Beeching rail closures era when it was one of the conditions for consultative committee approval of a planned rail closure. However, much earlier examples do exist. One such, and a fairly unique one, relates to the branch line from the Kings Cross to Peterborough main line at Holme, eastwards to the market town of Ramsey. Less than 6 miles long, this single line had been opened in 1863 and was one of a number of short branches linking adjacent population centres with the Great Northern Railway's main line.

Passenger traffic generated by Ramsey was limited by its modest size but, despite this, the town got a second terminal station when a second branch line was opened by the Great Eastern Railway (GER) from Somersham. At the beginning of the 1930s, the LNER had a serious

London & North Eastern Railway J6 Class locomotive No.3528 stands at the outer face of the Up platform at Holme at the head of what is either a local pick-up goods train or one destined for the Ramsey branch.

look at some of its poorly patronised lines and the ex-GER branch to Ramsey East lost its passenger trains in September 1830. In the following year a change was also made to the remaining Ramsey passenger facilities in the shape of an unusual alteration to the services between Holme and Ramsey North.

From 2 February 1931 onwards, the timetabled services between Holme and Ramsey became part train and part bus. The normal rail timetable was continued up to and including the 10.15 a.m. from Holme to Ramsey North, but after that the remaining four services of the day in each direction were replaced 'experimentally' by 'omnibuses of the Peterborough Electric Traction Company'.

It seems likely that this hybrid arrangement saved at least one train crew and may have avoided having to deploy a second locomotive to maintain the freight trips. Whatever the underlying reason, it seems to have worked well enough and continued for many years.

TWO FOR THE PRICE OF ONE

The Great Northern & Great Eastern Joint Railway line from March to Doncaster was built to provide a direct route between the Yorkshire coalfields and Eastern England. The Great Eastern Railway had sought this for years and finally achieved its objective as a result of an agreement with the Great Northern, which allowed the latter to protect its interests in the coal traffic by sharing in the new enterprise. The gestation of the new route had taken many years, but agreement was finally reached based on joint use of existing lines plus a new link between Spalding and Lincoln. The joint committee of the two railways became operational in 1882.

The 69½-mile route was primarily a freight line and was heavily used by coal trains, especially after the opening of Whitemoor marshalling yard in

1928. It was also used by some important through-passenger trains, but its local services, about four a day, could take over four hours to make their thirty stops.

The 'Joint Line', as it was always known, ran straight and flat through agricultural Lincolnshire, intent on its through traffic and a service to the bigger population centres at Spalding, Sleaford, Lincoln and Gainsborough. Intermediate communities tended to be small and scattered, but the timetable, at least, garnered in as many as possible to give the line one of the finest collection of double-barrelled station names of any. Its stopping trains served French Drove and Gedney Hill, Scopwick and Timberland, Blankney and Metheringham, Nocton and Dunston, Branston and Heighington and Haxey and Epworth.

The remoteness of some parts of the Joint Line was a headache for those called out in the night or bad weather to deal with emergencies. Even finding some of the places was a nightmare when faced with wandering country roads, a totally flat landscape and few lights anywhere. Another consequence of this isolation was that if a wagon was derailed it was often easier to leave it to rot (or be stripped by local action!) than to recover it for repair.

THE LONGEST TRAIN?

Really long trains have always been a rarity on Britain's railways. We have nothing to compare with the huge trains hauled by multiple power units on North American lines, for example. Seventy or eighty empty coal wagons could be seen on some main lines here, but train length was mostly restricted by block signalling considerations as much as by the availability of suitable motive power. A signalman working a fully interlocked frame could not pull off his signals if the far end of a train had not cleared a previous section.

Perhaps the most notable British example of a long train was the one booked to leave Penzance at 9.10 p.m. on Friday 20 May 1892. In May of that year, the Great Western Railway board had again discussed the matter of converting the company's remaining broad-gauge track to narrow, or standard gauge as it subsequently became known. The matter had been rumbling on ever since the appointment of the Gauge Commission back in 1846 and the recent Board of Trade requirement to improve signalling on the lines west of Exeter brought the final pressure that led to the GWR board's approval for the conversion. Fourteen votes were cast in favour of the change, only two against.

Thus it was that the last broad-gauge train from Paddington had set off for Penzance at 10.15 a.m. on that historic Friday. The whole system had been busy with getting men and materials on site for the physical work of conversion during the weekend and a great deal of preparatory work had been done.

Even so, train services had to continue to operate as usual during the Friday, although as the hours passed more and more stock was worked out of the far west in fourteen specials. Then it was the turn of the 9.10 p.m. which is reputed to have exceeded a hundred vehicles by the time it got to Swindon.

The last broad-gauge train had called at every station between Penzance and Exeter and every stationmaster had to be in attendance. He had to give his personal assurance to the inspector in charge that all broad-gauge stock had been cleared from the lines under his control and in return received a certificate confirming that the last broad-gauge train had passed. This certificate then provided the authority for the physical conversion work to start.

Incredibly, the huge conversion task was completed in time for narrow-gauge services to begin on the Monday. Over 250 miles of main line had had the third rail removed to change them from mixed to narrow gauge and 170 miles 63 chains of broad-gauge main line had also been converted.

RHUBARB! RHUBARB!

At one time, rhubarb carryings were an important source of railway revenue in some areas. There was even a 'rhubarb season' during which Kings Cross, for example, dealt with regular van loads of Yorkshire rhubarb every day. These had to be delivered early and in good condition to the London markets in order to fetch the best prices.

Rhubarb also gave its name to a railway bridge in Peterborough. When the Peterborough, Wisbech & Sutton Bridge Railway was built, ready for its opening in 1886, its exit route from Peterborough involved climbing to cross the Midland and the Great Northern lines on the north side of the city before heading off in a north-easterly direction. The soil used to create the embankment leading to this crossing, and an overbridge over the adjacent Lincoln Road, had previously been cultivated and turned out to contain a large number of healthy rhubarb crowns. These quickly took to their new environment and became so rampant that the road overbridge became known as Rhubarb Bridge.

TRANSITORY GLORY

Nicknaming railways is not a new practice. The London & North Western (L&NWR) delighted in its label as The Premier Line and the Great Western managed to go one better with its officially endorsed title of The Royal Road. Within the railway industry itself the references were rather less formal so that the L&NW was known to staff as simply the Wessie, the Great Northern was known as the Yorkie, the Great Eastern the Swedie, the Lancashire & Yorkshire the Lanky and so on.

In the British Rail period the line-naming activity spread wherever the marketing people could see the prospect of creating a favourable identity and securing the maximum media coverage. Thus were born the Wherry Line, the Tarka Line and dozens of others. The preservation activity kept pace with its Bluebell Railway, the Poppy Line, the Watercress Line and similar examples.

A warm smile and a bottle of Moet & Chandon champagne are part of the short-lived Heart Line's promotion of the Newcastle–Birmingham–Bristol route.

One such venture that hardly got off the ground was a scheme to publicise the important cross-country route from the far west, through Bristol and Birmingham to Sheffield, West Yorkshire, York itself and on northwards. This was to be the Heart Line to convey the stronger message that the service it provided ran through the heart of England.

The route now had High Speed Trains and quite a few of these received the Heart Line legend above the driving cab window. Hardly had this happened when the plan foundered with the creation of the train operating companies and the transfer of the route and services to one of the new franchises. A very short-lived 'line' indeed.

A SECOND'S REFLECTION

The influence of Brunel upon the Great Western Railway was, of course, immense. It might never have been completed without him. Equally, it might never have survived without his 'locomotive assistant' – to quote his job application – Daniel Gooch. In those early years Gooch showed his worth, especially in rescuing Brunel from his motive-power problems. Like Brunel he, too, was a 'hands on' man as his report on an incident in Box Tunnel shows.

Initially there was only a single line through the newly opened tunnel and, for the first forty-eight hours of services on the infant Great Western line, Gooch himself acted as pilotman on the trains passing through the tunnel, which had proved such a challenging and controversial enterprise. His presence averted a disaster which would have fuelled the public anxieties about travelling at speed through nearly 2 miles of the poorly lit bore. Gooch describes the event thus:

> At about 11 o'clock on the second night we had a very narrow escape from a fearful accident. I was going up the tunnel with the last up train when I fancied

> I saw some green lights in front. A second's reflection convinced me it was the Mail coming down. I lost no time in reversing the engine I was on and running back to Box Station with my train as quickly as I could, when the Mail came down behind me. The policeman at the top of the tunnel had made some blunder and sent the Mail on when it arrived there. Had the tunnel not been pretty clear of steam, we must have met in full career and the smash would have been fearful, cutting short my career also. But, as though mishaps never come alone, when I was taking my train up again, from some cause or other the engine got off the rails in the Tunnel, and I was detained there all night before I got straight again.

What the train's passengers thought of this to-ing and fro-ing is not recorded but, at the end of 'two days and nights pretty hard work', Gooch professed himself 'not sorry to get home and to bed'.

Apart from a year away to lay the first Atlantic cable using Brunel's great steamship *Great Eastern*, Gooch was to serve the Great Western Railway for the rest of his life. As its chairman, he steered it through some very bad times and continued in office until his death in 1889.

'LIKE AN EMPTY BEER BUTT ... '

As construction on the pioneer Liverpool & Manchester Railway (L&MR) neared completion, the thoughts of the directors turned towards methods of working their railway. With locomotive design very much in its infancy, the use of stationary engines was a strong possibility, but before any final decision was taken it was planned to invite locomotive engineers to take part in a competition to demonstrate what their machines could do. This led to the famous Rainhill Trials, in October 1829, following the publication of the competition invitation in the previous April. This had been

Novelty had been the favourite to win the Rainhill competition and impressed greatly with a speed of 28mph on the first day, but then faulty bellows ruined its chances. (Roy Gallop)

addressed to 'engineers and iron founders' and offered a premium of £500 for a machine that showed it could run forty times over a 1¾-mile distance at not less than 10mph and drawing three times its own weight. The forty runs stipulated would equate to a round trip between Liverpool and Manchester.

Thomas Brandreth, one of the L&MR supporters who had advocated the use of stationary engines, entered a horse-worked machine named *Cycloped* and another unlikely design utilising manpower, but neither was a serious contender. There were just four of those: *Rocket* entered by Robert Stephenson & Co., *Novelty* by John Braithwaite and John Ericsson, *Sans Pareil* by the Stockton & Darlington Railway's Timothy Hackworth and *Perseverance* by Timothy Burstall of Leith. There were three judges and, as history has so firmly recorded, *Rocket* won itself lasting fame by out-performing the other contenders.

One of the witnesses to the Rainhill drama was John Dixon who had taken over the task of being the Liverpool & Manchester's resident engineer and successfully resolved its construction problems, including that of crossing the notorious Chat Moss bog. In a letter to a friend describing the Rainhill events he proved himself a witty observer as the following extracts, from which the heading of this piece is taken, reveal:

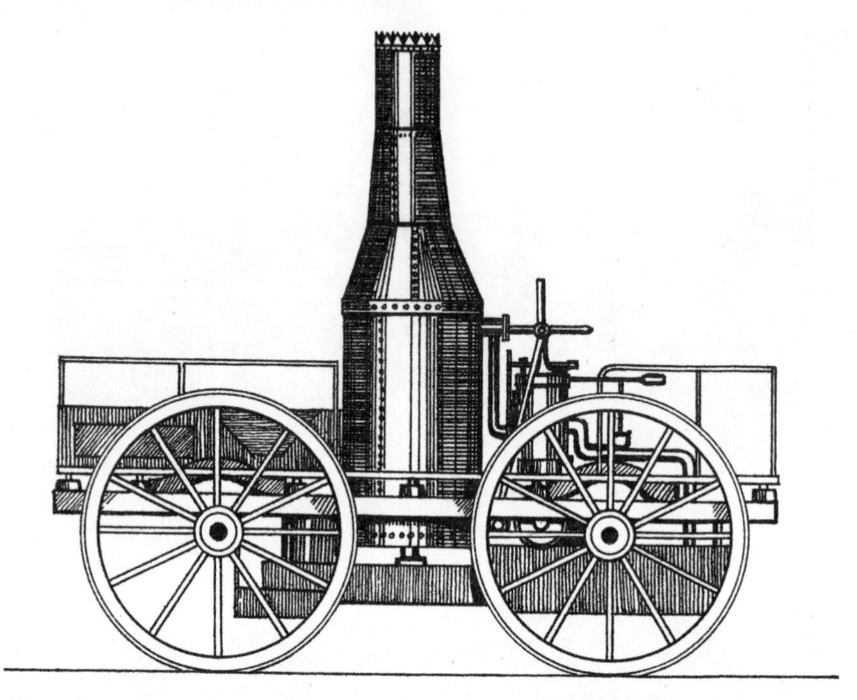

After a few runs and managing only a very low speed, Timothy Burstall had to give up any hope of winning the £500 Rainhill prize and withdraw his locomotive *Perseverance*. (Roy Gallop)

> The Rocket is by far the best Engine I have ever seen for Blood and Bone united. Timothy (Hackworth) has been sadly out of temper ever since he came for he has been grobbing on day and night and nothing our men did for him was right. He got many trials but never got his 70 miles done without stopping. He burns nearly double the quantity of coke that the Rocket does and mumbles and rolls about like an Empty Beer Butt on a Rough Pavement and moreover weighs above 4½ Tons. She (the Hackworth engine) is very ugly and the Boiler runs out very much, he had to feed her with more Meal and Malt Sprouts than would fatten a pig.

Dixon commented that the *Novelty*, a much lighter machine than the *Sans Pareil*, and the *Rocket*'s most serious rival, seemed to 'dart away like a Greyhound for a bit but every trial we had some mishap ... so it was no go.' His descriptive turn of phrase returned with full force when referring to Burstall's engine. This had unfortunately been involved in an accident on the way to Rainhill resulting in it being the last to start its trials. 'And a sorrowful start it was,' notes Dixon, 'full 6 miles an hour cranking away like an old Wickerwork pair of panniers on a cantering Cuddy Ass.'

WC&P

Like many light railways, the Weston, Clevedon & Portishead (WC&P) affair was idiosyncratic, to say the least. For a start, it took over thirty years from the original proposition to actually running its first train. The difficulties and setbacks of this period were legion, due largely to an on-going shortage of funds. One such problem related to the track where Scots pine had been brought down by boat to Clevedon to economise on providing sleepers, only to have a number discovered to have rotted by the time opening could be contemplated.

Eventually the whole route was operational, but it never earned enough to meet operating costs and service capital and the railway spent many years in receivership. One consequence of the need for economy was that much of the equipment was secondhand, resulting in a motley collection of locomotives, rolling stock and pretty well everything else. Most of the numerous halts were very simple affairs and when the one at Broadstone finally got a shelter, it was about the size of the average sentry box. To 'save money' the WC&P would haul wagons all the way from Clevedon to Portishead and back because they could be weighed there a few pence less than the local gasworks charged.

When Colonel H.F. Stephens added the WC&P to his stable of light railways he made some useful economies including the introduction of petrol-engined rail motors and tractors, but this also added to the incredibly mismatched appearance of most of the railway's passenger services. Again to save money, he decided to bring in Welsh coal by sailing vessel to a wharf on the River Yeo, adjacent to the main route of the railway and purchased the *Lily* for this purpose in 1927. Just under two years later, she sprang a leak on a journey over from Newport, soon proving impossible to steer and drifting helplessly west before the tide took her back towards her starting point. Instead of saving the stricken vessel, a tow from the Newport pilot cutter opened her seams and let in more water, causing the two-man crew to leap for their lives as the *Lily* turned turtle.

The WC&P was proud of its heaviest locomotive, the 2-4-0T *Hesperus,* but in 1934 she proved too weighty for a small bridge on the siding leading to the riverside wharf. Luckily only the timbers gave way and the rails just sagged, but the Great Western gave a dusty answer to a request for its breakdown crane, fearing it might suffer a similar fate on the bridge, which carried the WC&P main line over the river. Instead the locomotive had to be recovered by a laborious process of jacking and packing using the WC&P's own resources.

Much romanticised in more recent times for its undoubted character and probably quite useful to its local users, the WC&P finally gave up its struggle to exist in 1940, when wartime pressures for metal consumed its rails. Totally in keeping, subsequent thoughts of possible reopening of some part of the route have been thwarted by the total disappearance of all its title deeds, shareholding data and other documents. And when Radio Bristol conducted an interview with Christopher Redwood, author of a book on the line, things once more did not go as planned. The event was to take place on the banks of the River Yeo to give it 'atmosphere', but once there the rain came down so fast as to erase the interviewer's list of questions and an inquisitive herd of cows tried to nudge both parties into the river itself.

BRUNEL'S SPECTRES

The construction of the original GWR main line from London to Bristol was a mammoth undertaking for its day. It was not only challenging in engineering terms, but also produced a succession of accidents, natural hazards and other obstacles to be surmounted, which only the drive and energy evinced by Brunel managed to overcome. Despite the optimism he so often displayed, Brunel was not without his private anxieties and demons. The floods that threatened the final approach to Bristol must have seemed like the last straw and may have influenced his writing that:

> If I ever go mad, I shall have the ghost of the opening of the railway walking before me, or rather standing in front of me, holding out its hand, and when it steps forward, a little swarm of devils in the shape of leaky pickle-tanks, uncut timber, half-finished station houses, sinking embankments, broken screws, absent guard plates, unfinished drawings and sketches, will, quietly and quite as a matter of course and as if I ought to have expected it, lift up my ghost and put him a little further off than before.

NUTS

Higham Ferrers did not get a railway until 1893 and lost it again in 1969. Despite the strong representations of the local footwear manufacturers, the Midland Railway was seemingly uninterested in linking the town with the main line due to financial constraints and the opposition from vested land interests. Eventually, it was agreed to build a line through to Raunds on the Kettering–Huntingdon route, but this got no further than Higham Ferrers.

An unusual feature of this modest terminus was a prolific walnut tree in the station grounds. One of the obligations laid upon the Highham Ferrers stationmaster was to send 20lbs of walnuts every year to the Midland Grand Hotel at St Pancras.

RUNAWAY REAPER

No one was seriously hurt in the accident that took place on the Somerset & Dorset line in July 1936, but it was an extremely spectacular affair, worthy of a Will Hay or Carry On film sequence. It all took place on the section of railway between Radstock and Bath with the drama playing itself out over some 10 miles of undulating track and reaching its climax on the steep descending gradient on the approach to Bath Junction.

Radstock, with its two adjacent stations, lay in the heart of the North Somerset coalfield and had neighbouring sidings serving the collieries. There were always empties to be positioned and loaded wagons to be marshalled and so it was that a freight train headed north through the complex behind a 2-8-0 freight locomotive on this fateful day. Normality ended as it ran through signals at Writhlington and headed straight for an 0-6-0 tank shunting a few wagons at the sidings serving Braysdown colliery.

Realising what was about to happen, the crew of the freight engine leapt from the footplate and left their train to its own devices. The tank engine driver had also been keeping a good lookout and started to set back to minimise the force of the impending collision. He then realised that the speed of the main line train was slow enough for him to transfer between the two machines and bring the freight train to a halt. Unfortunately, he left his footplate without telling his fireman what he planned and the latter promptly jumped down from the other side of the cab to leave the shunting engine and its eight-wagon train to its own devices.

With its regulator still open and a couple of short down gradients to help, the now driverless runaway gathered speed through Wellow and accelerated even more on the 1 in 100 and 1 in 60 slopes approaching Midford. Derailment of the wagons was inevitable, but they mowed down the signal box in a resounding encounter before being catapulted down an embankment.

Leaving the wrecked signal box behind and its dazed signalman to contemplate the mess of derailed wagons and demolished lineside furniture, 0-6-0 tank engine No.7620 sped on, now with just half a wagon remaining of its original charges. The impetus was enough to get this bizarre ensemble up the rising gradient of the single line approach to Coombe Down Tunnel and down through Devonshire Tunnel on the other side, but the trail of havoc was nearing an end. An overbridge, coupled with an end door dropping off the wagon remnants, derailed it and the fusing of a firebox plug on the locomotive at the same time finally brought the destructive runaway to a halt.

THIRTEEN INTO FOUR

In the middle of the Second World War, a committee of twelve architects concerned with the reconstruction and modernisation of London's railway termini came up with their findings after two years of deliberation. Their report advocated a reduction to four main stations, no doubt to be architectural masterpieces. Paddington was to remain for services to and from the west, there would be a northern terminus embracing Euston, St Pancras and Kings Cross, an eastern one combining Liverpool Street and Broad Street and, most astonishing of all, a single southern terminus. The latter would, considered the committee, deal with all trains running to and from the traditional locations of Victoria, Waterloo, Charing Cross, Holborn Viaduct, Ludgate Hill, Cannon Street and London Bridge.

The source of this information does not mention Marylebone. Nor is any explanation given as to how this incredible transformation is to be achieved. One look at a map of the lines involved might have given the committee pause.

SIGNAL BOX NAMES

Signal box names are, for the most part, functional and related to the location of the installation. The most common supplements to a place name are the points of the compass: North, East, South and West.

Some places used all four of the compass group and on some journeys one might have passed all four. A passenger travelling from Spalding to Lincoln via Sleaford could have done so in the S.E.W.N. order and one from Ely to Peterborough with an order of S.E.N.W. at March.

Of course, there were more exotic and unusual signal box names such as Severus Junction near York named after the Roman emperor who died there in AD 212, and Bo-Peep Junction at St Leonards, derived from an adjacent hotel. Pouparts Junction, Whisker Hill Junction and Rose Heyworth North and South commemorate local people and places, as Chaloners Whin probably does. Point Pleasant for a signal box in the heart of Battersea seems something of a misnomer and Middlesbrough's Erasmus somewhat pretentious, even for a marshalling yard.

ABANDONED TRAIN

The Coronation Scot train was sent to the USA before the Second World War and did a tour of over 3,000 miles before being exhibited at the World Trade Fair in New York. The train crew returned to this country in 1939 and shipping space was subsequently found for the locomotive. But with every inch of cargo space needed for essential supplies, room could not be found for its eight-coach train and it had to be left behind.

Eventually the position had to be accepted and the London, Midland & Scottish Railway (LMS) decided that its high quality coaches must reluctantly be abandoned and some worthwhile use found for its valuable combination of seating, dining and sleeping vehicles. Accordingly, using the Baltimore & Ohio Railroad as it agent, a deed was drawn up transferring the eight coaches to the US government. That body, in turn, arranged for them to be stationed at Jeffersonville, Indiana and used as living quarters for the officers of the army's Quartermaster Corps.

THE DROP

Many early railway schemes suffered from conditions imposed upon their construction during the process of securing Parliamentary approval. None more so than the Edinburgh & Glasgow Railway (E&G), on which a great deal of money was expended with the intention of creating a fast, level route. Between Haymarket and Cowlairs the objective was achieved, but opposition from the Forth & Clyde Canal undid the good work by getting very onerous conditions imposed on the final stretch into Glasgow Queen Street station. Instead of dropping gently over the last few miles, the E&G

was forced to pass under the canal's Port Dundas extension by means of a steep gradient within a tunnel. This incline was over a mile long with gradients of 1 in 50 or worse.

It was beyond the braking capacity of early locomotives to descend such an incline safely and an imaginative solution for the problem had to be found. This took the form of building special all-metal brake vans equipped with extra powerful handbrakes and manned by special brakesmen, one for each van.

On the arrival of a train at Cowlairs, the locomotive would be detached and replaced by two or three of the special vans, depending on the size of the train. This took place while tickets were being collected and, at the same time, the train engine ran round its carriages and buffered up to the Edinburgh end to give the train a start down the incline. Under the control of the brakesmen it then descended to Queen Street.

For trains in the other direction a locomotive would descend the incline to take up position at the head of an outgoing train where it was then attached to an endless cable worked from a winding house at Cowlairs. Thus assisted, it would set of from the gloomy station confines and up the tunnel. Nearing the top it had sufficient speed to overtake the cable and the inverted coupling attaching cable to locomotive would fall away to allow the train to operate more normally.

Curiously, a similar situation existed at Edinburgh where the Edinburgh, Leith & Granton Railway used special brake vans on the section between Scotland Street station and the terminus. The latter was reached via a rising gradient, the train engine being detached and replaced by the winding cable and the extra braking power added to the rear, ready for the subsequent descent of an outward service. This took on its locomotive at Scotland Street and the brakesmen transferred to every second coach where they watched over the descent to the Firth of Forth from an exposed position with their head and shoulders above roof level.

THE INNOCENT RAILWAY

This was the name bestowed on the pioneer Edinburgh to Dalkeith Railway opened in 1831 from a depot near Arthur's Seat and descending through a tunnel forming part of a long gradient of 1 in 30, before heading east. The name derived from the line's accident-free record, but it was only horse-worked so speeds were not very great.

The initial section from the terminus was cable-worked with a rather unusual method of signalling installed. To start the cable movement use was made of a narrow metal tube running from the bottom of the approach incline to the engine house at St Leonards. By means of a set of bellows, air was pumped up the tube and used at the top to operate a bell and alert the engineer to start his winding engine.

TAUNTON BRAKE VAN

Attaching an extra brake van to heavy trains on steep gradients was not all that unusual, but Taunton had a special van used exclusively for this purpose. It was kept there for addition to extra-long passenger trains heading for Exeter, to guard against the consequences of a breakaway on Wellington Bank where the GWR main line involved a steep 4-mile climb to the summit beyond Whiteball Tunnel.

BRIEF GLORY

For most of its life the short Somerset & Dorset Railway branch, from Highbridge to Burnham-on-Sea, led a modest existence except on the warm days of summer when unusual numbers of holidaymakers and folk from central Somerset would arrive to enjoy the wide expanse of the resort's sands.

The tiny station's claim to distinction arose from the shipping link from the nearby pier across the Bristol Channel to Cardiff. Based on this, for nearly two years from May 1865, the railway advertised and operated an international 'Channel to Channel' service. Passengers from the Welsh capital could join the steamer at a quay in Cardiff Bay, use train services from Burnham to Poole, catch another steamer there for Cherbourg and eventually arrive, courtesy of the Western Railway of France, in Paris.

Burnham was an early victim of the Beeching rail closures but for a time, after normal services ended, a train continued on from Highbridge to the Burnham terminus on Wednesdays and Saturdays in summer. On at least one occasion when the rain poured down, a piece of local railway initiative arranged for the return service to depart earlier and hired a loudspeaker van to pass along the sea front offering this sanctuary to the town's wet and bedraggled visitors.

LONELY GIANT

One London & North East Railway (LNER) locomotive spent the whole of its working life in one location working backwards and forwards over the same 2½-mile stretch of railway. This was the company's only articulated engine, the Class U1 2-8-8-2 Beyer Garratt numbered 2395, later 9999. Of course, it also worked to and from shed and went for shopping but the giant's principal role in life was to help 1,000-ton trains of Yorkshire coal up the 1 in 40 section of the 7-mile Worsborough Bank climb to Penistone prior to the descent towards Lancashire.

After marshalling at Wath Yard, the coal trains usually set off with two 2-8-0s for the first part of the journey, but then at Wombwell one of these would be detached and moved to the rear where it was joined for the steep section ahead by No.2395. With this huge tractive effort the heavy load was hauled up the bank as far as West Silkstone Junction, where the Garratt came off and returned light engine to repeat the process again and again.

This impressive locomotive did get a change of scene after the Woodhead Tunnel route was electrified, but was still confined to a similar role, a similar round trip and a similar tough gradient, just up the Lickey Bank in a more rural Worcestershire setting.

THE GALLOPING SAUSAGE

An effective relationship with his chief general manager, Sir Ralph Wedgewood, enabled Sir Nigel Gresley, as the LNER's chief mechanical engineer, to enjoy a very innovative period once the 1923 Grouping Act had settled down. In 1927, he began thinking that a two-stage compounding system for using steam might be a way to reduce the massive bill his railway had to pay for its locomotive operation. A select group worked secretively on the new project over a period of some three years, which resulted in it attracting the label 'Hush-Hush', but it finally came to fruition with a streamlined 4-6-4 locomotive given the number 10000. It was not a good-looking machine as the fairing covering the water drums and tubes gave it a bulbous look. Nor did it perform as Gresley had hoped, for while it managed to handle heavy expresses well enough it proved prone to leaking tubes and other defects. Another problem was that maintenance costs proved high and the coal consumption economies disappointing.

Sadly No. 10000 could not live up to Gresley's expectations and, among the less understanding, its bulbous appearance led to the derisory label 'The Galloping Sausage'. Converted to a conventional three-cylinder form, the engine was relegated to normal Kings Cross–Doncaster workings with its only claim to fame that of being Britain's only 4-6-4 locomotive.

18HP

Devon's first railway was the Haytor Granite Tramway, opened in 1820 to convey granite from quarries on the edge of Dartmoor down to the Stover Canal for onward movement by barge to Teignmouth and shipment from there. The wheels of the tramway waggons had no flanges for they ran on shaped sections of granite, each up to 8ft long.

Horses were used to pull the twelve-waggon trains, no less than eighteen of them harnessed in single file behind to restrain the descent and in front for the upward climb with the empties. This 4ft 3in-gauge line lasted until 1858.

TWO UNKNOWN

These pitiful words on a tomb in South Gloucestershire represent the only memorial to two unnamed children who died in a terrible accident at Charfield on the main line from Birmingham to Bristol on 13 October 1928. On the foggy morning of that sad day, an express passenger train from Newcastle crashed into a goods train in the process of setting back into a siding to clear the main line. The scattered wreckage was then hit by another goods train heading north on the opposite line. Gas cylinders used to supply the lights in the carriages of the express exploded, resulting in a searing blast and a fierce fire. Many were trapped in the wreckage.

Painstakingly the bodies of the dead were examined and identified, except for the two youngsters who, tragically, came to be called the 'two unknown' and were only remembered as such in a country graveyard.

NOT JUST LEAVES

Train travellers in Britain understand that decaying leaves, like ice, can seriously delay trains by affecting traction adhesion and current supply. But imagine their reaction if the now-familiar announcement changed to, 'We are sorry to announce that your train will be delayed due to caterpillars on the line.'

This would not have been a ridiculous scenario on the Cumbres & Toltec Scenic Railroad in the United States; a narrow-gauge tourist line linking Colorado and New Mexico. Apparently millions of the creatures produced an unusual level of activity which saw them swarming in the hilly area through which the line passes and covering its tracks with their slimy deposit. It took steam spraying and some double heading to keep the trains moving.

THIRTY-SEVEN MINUTES' DELAY

For embryo railway routes having to cross a wide coastal estuary was a nightmare, not only in construction and cost terms, but also in terms of operational constraints. Until the second half of the nineteenth century, sailing vessels outnumbered steamers and their tall masts meant that water crossings had either to be built high or with an opening section. The combined weight of the Admiralty and the Board of Trade reinforced this constraint.

Eastern Britain favoured swing bridges but further west a retracting section was sometimes used. At Bridgwater, for example, an extension from the Somerset & Dorset station to the west bank of the River Parrett

The view south across the Afon Mawddach showing the railway bridge linking Barmouth with the coastal route on to Fairbourne and Tywyn.

utilised a crossing in which a portion of the approach track moved sideways to allow the length over the river to slide back in its place.

The three estuaries of the west coast of Central Wales presented a real challenge for the piecemeal railway construction there. At Aberdovey and Portmadoc the route looped inland to shorten the crossing but that across the Mawddach Estuary used a bridge to replace a really horrendous two-boat ferry predecessor. It took some building but the 121-span structure was eventually completed in 1867, but then failed to pass inspection by the Board of Trade's Captain Tyler. He had reservations about some details of the construction and declared unacceptable the fact that it took two men no less than thirty-seven minutes to open the retracting section and then close it again. Later railway operators would have agreed wholeheartedly and sailing vessels' masters trying to hold their vessels against a strong tide while they waited to pass through the opening span may well have felt the same way.

ON THE LEVEL

'All Trains must stop at the Water Tank on the Burnham side of the Bristol and Exeter Line'. This instruction and a further one limiting speed to 4mph appeared in the Somerset & Dorset Railway Rule Book and referred to the ¼-mile section of railway from Highbridge B signal box to the GWR Highbridge West box at the Bristol end of the GWR's Down platform. The latter box controlled the point at which the Somerset & Dorset Railway's Somerset Central west–east route from Burnham-on-Sea to Evercreech Junction crossed that of the north–south Bristol to Exeter line.

Railway routes crossing one another at right angles or obliquely, and without connection, were rare and appeared to invite disaster. But they were a good way of avoiding the high cost of bridging in a flat landscape and, with signalling control vested in one signal box and suitable interlocking, were perfectly safe.

The track layout at Newcastle Central was notable for its complexity and the number of diamond crossings. The signal gantry was quite impressive too.

The arrangement did make life a bit more complicated for the Timing & Diagramming office people in the days when train paths were worked out on a huge sheet with the route along one side and the hours of the day along another. Squeezing in the short-line occupancy that a flat crossing entailed could get really frustrating, and nowhere more so than in relation to the East Coast Main Line (ECML) which was unique in having no less than three of these crossings; located at Newark, Retford and Darlington.

There was another flat crossing of two rail routes at Murrow where the modest train service between Peterborough and South Lynn had to cross the unending stream of coal and other freight trains running along the Great Northern & Great Eastern Joint Line from Doncaster to Whitemoor marshalling yard. A further flat crossing was located just outside Hull Paragon where the North Eastern Railway route to Scarborough intersected the connection between the branches diverging left towards the riverside wharves and right to the route across the River Hull to the main docks and the Hornsea and Withernsea branches. Less formal crossings were not unusual in dock, factory and colliery sidings but nowhere could match in stature the three on the ECML.

Another rare unlinked route crossing, this time of the Midland & Great Northern and Great Northern & Great Eastern Joint lines at Murrow. (G. Orbell)

Closures, diversions and the building of an underpass at Retford have all relegated these flat crossings from their former status. They no longer cause regulation disputes where, as in the Newark case, the trains on one line had a different control office to those on the other and both had their own priorities.

SIDMOUTH ROAD

Rail passengers learned to be wary of station names which embraced the word 'Road'. It invariably meant that the place in the preceding name was anything but nearby.

Before the era in which railways seemed to intend to serve everywhere it was expected that only the wealthy would use them and that intending travellers would get to their nearest station by coach or post chaise, so that including 'Road' in a station name was perfectly sensible. Indeed stage coach operators proved quick to adjust their activities to provide feeder services to their new rivals.

An outstanding example of the 'Road' syndrome appeared in a Bristol & Exeter Railway Board minute that planned to call a new station on the approach to Exeter 'Sidmouth Road'. In fact its location, beside the River Culm at Hele, was some 14 miles from Sidmouth!

In fairness, the Bristol & Exeter Railway directors had been influenced by a proposal emanating in Ottery St Mary to build a turnpike road between the new station and the growing seaside town. This, however, came to nothing and the route to Exeter was opened on 1 May 1844 with Hele having just a simple station and a simpler, more appropriate name.

'KIM' SAVES A TRAIN

There are many examples of dogs collecting money for charity on railway stations, but not too many of them acting as permanent way staff. One example did, however, appear under the above heading in *The Outline* in September 1928 where the account read as follows:

> This is the true story of an Airedale.
>
> Part of his master's duty was to see that the railway line between Mora and Mallaig was kept clear of landslips, the hillside between these two places being very steep and broken at parts, and landslips were not uncommon at certain seasons.
>
> One evening, when the nights were closing in, his master returned as usual from his evening's survey – all was well, and he was enjoying his well-earned and comforting tea when Kim dashed into the kitchen in great excitement and wouldn't be satisfied until his master again faced a cold and inclement outing. On he went, with an occasional glance at his master as much as to say 'Now don't turn back, just wait and see', and sure enough, in the darkness before him, a landslip had covered the lines that an hour before had been clear; and the evening train was on its way; would he have time to signal the 6 o'clock passenger train before it rushed to destruction? Back rushed man and dog and were just in time to put up the danger signal before the train came in sight.

TAKING THE BISCUIT

The biscuit firm Huntley & Palmer was a major employer at Reading and had extensive factory premises beside the adjacent GWR main line with a network of sidings for loading their products to freight vans. Such was the volume of business sent away by rail that the tins were specially shaped to secure the best possible loading in the wagons and one outwards freight service was even officially known as 'The Biscuits'.

To advertise their products Huntley & Palmer at one period were distributing biscuits free on Reading station to passengers bound for London. Only those in first class, mind!

HARBOUR OF REFUGE RAILWAY

For some sixty years herring fishing was a major industry along the east coast of Britain. The herring fleets started the season at Fraserburgh and Peterhead and then followed the great shoals of fish down to Yarmouth and Lowestoft, where thousands of tons of herring were loaded onto to rail. At Peterhead the fishing activity expanded rapidly in the 1880s, quickly outgrowing the modest facilities there and creating a need for an improved, all-weather harbour.

To build a new breakwater, the Admiralty needed a good supply of stone and a granite quarry at Stirling Hill, a little over 2 miles inland, was well placed to supply this. Labour was also needed and the prison which was opened in 1888 at Peterhead represented a good source of manual workers at very little cost. All that was needed was a way of linking the two places. Accordingly a short single line railway was built from Peterhead out to the quarry.

The Harbour of Refuge Railway, as this line was known, was built to high standards and had its own locomotives and rolling stock. The quarry now had a captive labour market and an easy way to carry workers to and fro. Special vans were provided, with secure compartments and designed to hold up to thirty-five prisoners, shackled and watched over by prison guards. When the line closed, several of the 1915 build of coaches found a new home on local farms. Happily an example of one of the vans still survives although its unusual line has long since disappeared, as has Peterhead's own 13-mile Great North of Scotland link from Maud Junction.

BOVINE BALLAST

The Listowell & Ballybunion Railway was a curious line of 10 miles 20 chains that linked Listowell with the smaller town of Ballybunion and the waters of the Atlantic Ocean near the mouth of the River Shannon. Opened in 1888, its track was built on the Lartigue system of metal trestles topped

The curious twin-boilered locomotive of the Listowell & Ballybunion Railway, built for them by the Hunslet Engine Company.

by a single running rail, which the twin-boilered locomotive and its double compartment coaches straddled. Acting as canal drawbridges, seventeen 'Flying Gates' served as crossings, with keys for them being held by the line-side farmers.

To keep the rolling stock balanced, passengers had to be seated in each of the twin compartments. Freight represented more difficulty and on one occasion a cow had to be borrowed to balance the piano that had been consigned. There was then the problem of returning the cow. A novel solution was found; counterbalancing the cow with two calves who themselves got home by being placed one on each side of their wagon.

Despite its peculiar nature the line was capable of earning more than its running costs but not by a sufficient margin to fund repairs and renewals and it finally had to close in 1924.

I SEE NO SHIPS

Along the single line from Bristol to Portishead a complex of station, passing loop, signal box, sidings and branch line was constructed at a point 127 miles 30 chains from London, but never fulfilled the purpose for which it was intended. The fifty-seven-lever signal box was brought into use on 29 January 1918 and lasted until 14 April 1964, by which time the station itself had been closed for twenty-one years.

This situation came about as a result of the considerable losses of merchant shipping suffered during the First World War. To maintain the country's vital supply lines it was decided to build a new shipyard on the south bank of the River Avon, not far from its emergence into the Severn Estuary. A new station and associated sidings on the Portishead branch were linked to the shipyard construction site by a connecting line and were, for a short time, used for bringing in the materials for the building work. But then the war ended, the need for the additional shipbuilding capacity no longer existed and the associated part of the new station facilities became redundant.

The station, named Portbury Shipyard, was in use from 16 September 1918 to 26 June 1923 and had been intended for the use of people working at the shipyard site but that need passed when the government abandoned the project. The GWR retained the signal box and passing loop but, apart from the original construction materials and a few loads of coal, the sidings and shipyard branch had become redundant and were eventually lifted.

UNCOMMON CARRIER

Passed to deter railways from abusing their growing commercial power, the 1838 Carriers Act obliged them to accept and carry pretty well anything that was not harmful. This occasionally led them into some bizarre situations, one of which developed in East Anglia once the main line railway link from Yarmouth to London became available from 1845.

The background was the rapid growth of medical knowledge, stimulated by the consequences of war and the proliferation of medical schools. Under eighteenth-century law the essential supply of bodies for dissection had been met by the right to use those of executed felons, but the growth of transportation as an alternative had resulted in the demand for corpses exceeding supply and the horrid practice of body snatching becoming more prevalent. One location where resurrectionists operated extensively was the graveyard of the minster church of St Nicholas in Great Yarmouth. A notorious case there involved a group living nearby, in an alley which became known as Snatchbody Row, and who took bodies from the churchyard, carried them in sacks to a stable, loaded them into

The vast graveyard of St Nicholas Church at Great Yarmouth, once the haunt of body snatchers. (W.E. Parker)

ready-made boxes and then had a local carrier transport them to a waiting surgeon client in London.

Another statute, the Anatomy Act of 1832, passed as a result of the public outcry against body snatchers, made it legal to use the unclaimed bodies of paupers for dissection purposes. At first the workhouses could not meet the demand, but by the time Great Yarmouth had railway facilities the old resurrectionist trade had petered out and its practitioners had found new legitimacy in a different form. They formed a link with a Cambridge surgeon badly in need of cadavers for his students, and who duly contracted with shady East Anglian suppliers to obtain a regular intake by rail.

The Norfolk Railway agent at Great Yarmouth must have been surprised to be asked to send regular consignments of bodies to Cambridge every week, but the Carriers Act and the company's need for revenue meant that he was in no position to refuse. A tonnage rate was agreed, a special van and service arranged and up to a dozen bodies a week were soon making their last journey by rail. Out of these shady beginnings the railway network later become a legitimate and caring way of transferring the dead to their final resting place, notably expressed as an element in the state funerals of a number of honoured citizens.

OUT OF GAUGE

Although road haulage entered the business of carrying large loads in the 1920s using Scammell tractors and purpose-built trailers, railways continued to play a major part in this section of the transport business until the 1960s. It was a complicated and specialised area with no two jobs normally alike and railway movement severely hampered by the fact that each of the original railway companies tended to have its own loading gauge. This meant that the largest item that could be carried without special arrangements was one that could pass over the weakest track and beneath the lowest bridge or tunnel it would encounter on its journey. Each railway goods yard had a suspended, shaped metal bar over one of its sidings to mark the upper profile limit for normal conveyance.

The numerous loads outwith the normal loading gauge needed special arrangements to ensure that they did not foul bridges, stations, tunnels and signals and this could often mean using a circuitous route, sometimes even the closure of adjacent lines or running on the wrong line, all of which had to be carefully planned in advance.

Among the many 'out-of-gauge' loads carried were two such examples emanating from Chepstow in 1934. One of these involved a trainload of twenty-five huge Admiralty buoys destined for Newport Docks for shipment. Each buoy was 10ft high and weighed around 7 tons and they were loaded in pairs on a special train of Weltrol wagons. This movement needed special arrangements because of the width and overhang of the load, but an even tougher challenge came in the form of two massive bridge girders, destined for Bath and weighing 138 tons in total. For their journey from Fairfields works, each had to be suspended between two eight-wheeled Pollen B wagons, but their great length and total rigidity meant that every bit of track curvature was a risk that had to be assessed and overcome.

Despite the growth in heavy haulage by road, BR continued to play a part in this field for many years. In 1961, for example, it conveyed a massive transformer along the Conway Valley branch to Blaenau Ffestiniog. Both profile and weight were problems with this job. The former was overcome by building a dummy outline of the load on a conventional van and the latter by running a second locomotive behind the special train. If the dummy past the route features safely, so would the main load, and if there was a breakaway on a gradient, the following locomotive would act as a buffer!

THE SOUNDING ARCH

Unsurprisingly, in view of the size and novelty of the undertaking, there were divided opinions and conflicting factions within the upper echelons of the infant Great Western Railway. The Bristol Committee had more fanciful ideas about the appearance of the works than the London Committee, for example, while the Liverpool investors had serious doubts about some of Brunel's ideas. These doubts were reinforced when the 1837 opening of the first section of the line exposed shortcomings in both the locomotives and the track. The various internal conflicts continued and then surfaced again over the crossing of the River Thames at Maidenhead.

To avoid impeding the waterway traffic without resorting to severe approach gradients, Brunel opted for two low main spans supported on a pier rising from a shoal in the centre of the river. This involved daringly flat arches, each of the two main ones rising only 24¼ft above the river despite being 128ft long.

Construction work on the Maidenhead Bridge progressed well enough for the centring beneath the arches to be eased in the spring of 1838, but this immediately revealed some minor distortion in the eastern one. Like vultures, the Liverpool party insisted on their nominee inspecting the work but it was clear that the distortion was entirely due to the contractor easing the centring before the cement had set properly. Brunel insisted that he should put this right and the required work was duly carried out.

Brunel's critics were not entirely silenced and declared that the arch would surely collapse when the centring was finally removed. So Brunel left it in place, but must have been secretly delighted when a severe storm in 1839 blew the wooden support structure down and hurried it off ¼ mile down the river. And so the controversial arch stands to this day and derives its nickname from the notable echo it can offer anyone passing beneath.

A RARE SURVIVOR

At one time there were many lonely and isolated stations in Britain, especially in Wales and the Highlands of Scotland. Few have survived the years of 'rationalisation' but one that has is located, surprisingly, in Norfolk. This station, actually just a very short and simple platform, lies 3¼ miles along the nearly straight 8 miles of single line between Reedham and Great Yarmouth.

The first railway in Norfolk, the Yarmouth & Norwich Railway (Y&NR), was authorised in June 1842 and duly opened to the public just two years later. It took advantage of the flat landscape along the route of the River Yare, leaving Yarmouth to follow the edge of Breydon Water and then heading through Acle Marsh and Reedham Marsh to Reedham itself, where it now meets the line from Lowestoft, before heading on via Brundall to Norwich. In this Up direction the first 8 miles runs through flat, waterlogged ground and had to be laid on a special bed of faggots and ballast to produce a firm foundation. It is a landscape without roads and with few buildings apart from the odd marsh-dweller's cottage or farmhouse and occasional water mills, now much reduced in number. For centuries it was brought to life by cattle being fattened there in the summer, but in winter it has always been a stark, lonely place, the domain of wildfowl more than of ordinary mortals.

The Berney family, from which the station and its own bit of marsh get their name, owned land here for hundreds of years and Thomas Trench Berney sold the Y&NR the portion needed for the building of the line. It was a condition of the sale that a station be provided near where the Yare and Waveney rivers meet to enter Breydon. But its patronage was, understandably in view of the character of the area, sparse, and the railway board decided to stop its trains calling there in 1850, arguing that the station had duly been provided as per agreement, but there was no obligation to actually serve it. Thus began ten years of litigation although the railway did restore services halfway through and a settlement eventually gave the

The tiny, isolated platform of Berney Arms, between Reedham and Great Yarmouth. (W.E. Parker)

A train leaves Berney Arms for Great Yarmouth through the flat, lonely marshes beside the River Yare. (W.E. Parker)

Berney side compensation of £200 and the right to have one train each way stop on Mondays, Wednesdays and Saturdays!

At one time thirteen trains called daily at the tiny Berney Arms platform, if requested, although this has steadily reduced to two each way, with more on Sundays. The Yarmouth postman used to come in by train weekly and collect the outgoing mail from the postmistress-cum-station mistress housed in one of the nearby railway cottages. It was even a block post with simple signals and Tyers token working as an aid to moving the many crowded holiday trains heading via Norwich to and from the seaside. In summer the Berney Arms near the river attracts passing trade from the Broads holiday craft, visitors to the RSPB bird and nature sanctuary and users of the two major walking routes. Some of these folk use the station, pushing the numbers up to well over a thousand a year, but winter is a very different story.

The closure-era criterion of available alternative road services had no bearing on the Berney Arms platform. The nearest road is 3½ miles away via a trackway across Halvergate Marsh. And so this rare little station survives with diesel multiple units bringing and collecting the occasional passengers and main line trains thundering past on the busiest of summer Saturdays.

STIFFS' EXPRESS

Although no outward irreverence would have been shown, the railway staff involved did use this impolitic but descriptive term for the movement of dead bodies by train from central London to cemeteries well out of the Metropolis. The situation arose from the rapid expansion of London in the first half of the nineteenth century and a series of cholera outbreaks resulting in the demand for burials greatly exceeding the space available in the inner graveyards. As a result, coffins were piled into what little space there was at the London churches, sometimes in what were just communal pits and frequently, especially in the case of paupers, with indecent haste.

This situation and an accompanying outcry led to the setting up of two enterprises to acquire land well out in the country for use as new cemeteries and an arrangement with the appropriate railway to run trains there as part of the funeral process. The first of these schemes crystallised in 1852 when the London Necropolis & Mausoleum Company was incorporated. By arrangement with the London & South Western Railway, a daily train left a special station adjacent to Waterloo at 11.35 a.m. to convey coffins and mourners the 28 miles along the main line to the newly created cemetery. Near what was later to be Brookwood station the train reversed onto a short branch line leading to the two cemetery stations, one for Anglicans and the other for various Noncomformist denominations. After the funeral, mourners were returned to London at around 2.15 p.m.

There were three funeral 'packages' on offer, the highest priced one offering an exclusive site at the cemetery and costing 6*s* for the first-class travel ticket and a pound for the conveyance of the coffin. On the train the lavishness of the accommodation differed according to class. On a busy day as many as 150 mourners might be carried, but on others when insufficient business was on offer, the train did not operate.

Although the Necropolis Company averaged well over 2,000 bookings a year, and did quite well out of a period of the mass exhumation of bodies in London to make way for new buildings and works, it was not really a financial success. This was due, in part, to the fact that its travel fares became too cheap in comparison with those of the main line railway. The end came in April 1941 when enemy bombing badly damaged the Waterloo station leaving only the receiving office building still to be seen together with a monastery on the site of the South station at the Brookwood cemetery.

In July 1861 the Great Northern London Cemetery began to fulfil the same function as its Brookwood counterpart. The London station, quite an elaborate affair, was located beside the main line just north of Gasworks Tunnel and the Great Northern Railway operated a train from there to a new cemetery at Colney Hatch, later New Southgate. For some years this cemetery was busy with burials, but the train service did not live up to expectations and ended quite quickly with the last service operating in April 1863.

WHAT'S IN A NAME?

Some railway companies had very tidy locomotive naming practices, the Great Western in particular. The London, Midland & Scottish Railway was also fairly tidy considering the size of its engine stud, with orderly groups of places, people, military connections and the like plus that huge list of names given to the Stanier Jubilee 4-6-0s – places all over the world and naval heroes, vessels and battles mostly.

For sheer and rather odd variety, the London & North Eastern Railway (LNER) undoubtedly took pride of place. The practice of giving the Gresley A3 Pacifics racehorse names produced a odd range from No.60038 *Firdaussi* to No.60055 *Woolwinder* and No.60075 *St Frusquin*. *Pretty Polly* for one of these magnificent power houses was not really very apt. The various A2 4-6-2 design rebuilds also enjoyed a naming variety including exotic examples like No.60514 *Chamossaire* and No.60539 *Bronzino*. Thompson's own contemporary design, the B1 class of 4-6-0s included forty engines named after animals like No.61012 *Puku* and No.61037 *Jairou*, with a dozen or so others named after people and the rest, apart from No.61379 *Mayflower*, carrying no such badge of identity.

The most colourful and idiosyncratic locomotive names were those reflecting the LNER's Scottish connection. Class A1 No.60143 was named *Sir Walter Scott* and a host of other locomotives bore names reflecting the influence of his writings. Rather strange to southern eyes and ears were *Jingling Geordie* and *Luckie Mucklebackit*, while the D11 4-4-0s No.62671 *Bailie MacWheeble* and No.62691 *Laird of Balmawhapple* were not much better. *Cuddy Headrigg*, *Wandering Willie* and *Dumbledykes* all added to this unusual group. *The Fiery Cross* struck a rather different note and the mighty No.60506 *Wolf of Badenoch* was as thrilling to behold as the original chieftain had been – an independent and fearsome thorn in the side of the High Kings of Scotland.

Platform Clutter

A very unusual piece of platform 'furniture' is this monster contraption used for cleaning the lights and high windows at Bristol Temple Meads.

Platform Clutter: At one time barrows for parcels and for passengers' luggage (here and previous page) were a common sight but must have taken a lot of handling. Behind the four-wheeled pair is a standard, highly decorated weighing machine. Another barrow variation stands between the two hand-pumped water bowsers.

Cross-platform Links

Cross-platform Links: An unusual feature at Brockenhurst was a pivoting bridge for the transfer of parcels across a running line (above), while at Halesworth on the East Suffolk Line (below) the level crossing included a section of the Up and Down platforms before the road was diverted.

Fanciful and Functional

Fanciful and Functional: The incredibly ornate Midland Hotel at Manchester contrasts with the downright ugly Great Northern goods warehouse at Kings Cross.

Little and Large

Little and Large: The best the impoverished Weston, Clevedon & Portishead Light Railway could manage to provide lifting capacity at its depot at Clevedon was this sheer legs structure, almost certainly homemade. It makes a marked contrast with the posed picture of two GWR giant cranes lifting *Dartmouth Castle* as if it weighed nothing.

Lofty Perches

Lofty Perches: Not all signal boxes fulfilled the popular line-side image. Some had to be fitted into very limited spaces, one at Bristol even clinging to a cutting wall. Here the London Midland box at Gloucester (top) and the nicely painted structure at Hexham (Roy Gallop) had the twin advantages of elevation and an excellent all-round view.

LOCOMOTIVE HYBRIDS

The firm of Aveling & Porter grew from the partnership of Thomas Aveling and Richard Thomas Porter at Rochester in 1862 to become the largest manufacturer of steamrollers in the world. The production of traction engines, ploughing engines and road rollers began in the Invicta Works at Strood and expanded over the years to include an odd-looking adaption of the basic design for railway use.

Early clients for the hybrid, with its traditional cylinder, flywheel and driver's position perched above a long-slung boiler and firebox and carried on rather small flanged wheels, were the local quarries, dockyards and then brick and cement works. The Lodge Hill & Upnor Railway took several of these machines and Brassey used them in the building of the East London Railway.

When it opened in 1871–72, the Brill Tramway branch from Quainton Road initially used horses but soon ordered two of the Aveling & Porter

Aveling & Porter locomotive Works No.9449 was originally purchased by the Holborough Cement Company, but after its industrial life ended it passed to the Bluebell Railway and is pictured there in August 1966.

0-4-0 geared 'locomotives' (Works Numbers 807 and 846) and paid £800 for them. They were replaced by Bagnall saddle tanks in 1876–77 but, after its later years with the Metropolitan Railway and the London Passenger Transport Board, the Brill line eventually closed in 1935.

Several examples of the Aveling Porter railway 'locomotive activity' have survived at preservation centres and heritage locations.

MURDER AND BEDLAM

A modest Somerset single line branch a few miles west of Frome seems hardly the location to warrant such dramatic terms. When the North Somerset line from Bristol to Frome closed, a portion was retained to serve Whatley Quarry and a wagon works at Radstock. Subsequently, the latter too succumbed but the portion north from Frome and then over the quarry's own line survived to become an important route for the forwarding of heavy trains of limestone hauled by the powerful Class 59 locomotives built by General Motors.

ARC's huge Whatley Quarry originally used a 2ft gauge system to move the quarried stone to Hapsford for transfer to standard-gauge wagons but the new line permitted standard-gauge stock to work through to the quarry with the help of Sentinel steam shunting locomotives. In due course, these gave way to diesel units with Class 59s then taking 4,000-ton trains along the surviving route to Frome, Westbury and the West of England main line.

On its way from the quarry to the BR system the access line runs through three tunnels, the 55-yard Murdercombe Tunnel, named after a local crime, the 319-yard Great Elm Tunnel and then the 275-yard Bedlam Tunnel. On a grey, damp winter day the route through dense scrubland seems to be well named.

FILLING STATION

In a surprising example of commercial enterprise, one of the GWR's smaller stations was able to retail petrol. This was Pilning, where the High Level was the last station before entering the Severn Tunnel from the London direction, and the date was 1921.

Railways had carried road vehicles since the opening of the London & Manchester Railway and seemed only too happy to carry the carriages of the wealthy almost on demand. The London & Southampton had even carried stagecoaches. The relationship between road and rail began to change with the arrival of the motorcar but, at that time, the early motorists between Bristol and South Wales had no means of crossing the River Severn south of Gloucester unless they used the small ferry which operated from Aust to Beachley.

A Great Western Railway 2-6-0 bursts out of the eastern end of the Severn Tunnel with a train of gas-lit clerestory coaches.

The GWR saw this as an opportunity to arrange a car-carrying service through the Severn Tunnel between Pilning on the Gloucestershire side and Severn Tunnel Junction on the Welsh side and had one up and running before the First World War. In 1921, the *Great Western Railway Magazine* reported the resumption of the service with 'cars now conveyed by any ordinary train booked to call at Patchway, Pilning and Severn Tunnel Junction.' The next stage was the operation of dedicated trains, usually comprising a tank locomotive, single passenger coach and a few four-wheel flat wagons, loaded via an end dock. Petrol tanks had to be emptied, but at the end of the journey the company not only replaced the fuel but expressed itself willing to sell motorists up to 6 gallons in addition to the amount that had been discharged.

A special rate of 7*s* 6*d* per car was announced in 1928 to encourage more business with two trains each way daily, then representing the basic service right up to the 1960s. The old single-vehicle wagons were replaced by bogie Carflats created from old coaching stock and the facility was quite well used, especially on Mondays and Saturdays. Inevitably, though, the advent of the M4 motorway brought about the end of both this rather unique facility and its ferry competitor.

STRANGE LOCOMOTIVE, STRANGE LOCATION

The strange locomotive was a horse-powered machine invented in Italy and brought over for trials at Nine Elms in 1850. It bore the name *Impulsoria* and performed quite creditably in comparison with the steam machines of the time.

Impulsoria was powered by either two or four horses walking at a consistent pace on a treadmill. Not a new idea, but the gearing fitted allowed the 'driver' to adapt the output from the horses to the track gradient and load by moving between cogwheels of differing sizes on the chain drive between the treadmill and the locomotive driving wheels. All mounted on an open six-wheeled frame, the machine was managed by two men, one mounted on the rear horse to control their activity and the other on the rear platform to deal with the gears and brake.

A twentieth-century reminder in Somerset of a machine that was tested at Nine Elms in 1850.

During its trials, *Impulsoria* achieved 7mph with a load on a gradient and was expected to be capable of more than twice that speed and to handle fifteen or more wagons. Running over 200 miles a day for a few shillings was also in prospect.

The Italian machine was put on show at the Great Exhibition but then disappeared from the home railway scene. Curiously, though, it was until relatively recently remembered on an inn sign adjacent to a bridge a short distance along the GWR branch line from Yatton to Clevedon in Somerset. Quite what connection there was between the 1850s events and this pub deep in Great Western territory is not clear, but it is hard to envisage modern trains hurtling along anywhere behind a horse-powered locomotive, even on the modest route of the former Clevedon branch.

BLENDING IN

There are three major banks on the main line from Plymouth to Newton Abbot. The first climb eastbound is the 1 in 42 up to Hemerdon summit. When the 1893 signal box there was replaced by a new one, the local farmer insisted on the timbers of the rear side being painted green to blend in with the local scenery.

LAIRA RAIDERS

Hemerdon Summit also featured in another unusual, and in this case rather reprehensible, railway incident. In a piece of local enterprise Laira loco depot at Plymouth was known to organise an engine and a couple of brake vans just before Christmas for an illicit trip to the top of Hemerdon Summit. 'Diked' in the loop there, the brake vans disgorged their complement of marauding engineers who dropped down to track level and then nipped over the line-side fence. Carrying saws from the depot's tool store, the party quickly 'liberated' a dozen or so Christmas trees from the large patch being grown there and hurried them back onto the train, and delivered them to the depot as well as to the firesides of a number of festive Christmases.

ALL THE WHEELS?

At the official enquiry into the 1983 sleeper train derailment in Paddington station, one of the witnesses was the driver of a passing London Transport train. When the proceedings came to consider his evidence he was asked about the speeding sleeper's braking. In reply he mentioned the sparks he has seen from the train wheel. 'Could you see all the train wheels?' he was asked. 'Only on my side,' was the response.

NOT TO BE DESPISED

Despite operating some 1,316 miles of railway, the Great Eastern Railway never paid above 2 per cent on its ordinary stock, often considerably less. Even so, its coverage of agricultural East Anglia was important and extensive. For example, one modest station on one modest branch line, Cold Norton on the 8¾-mile Essex branch from Woodham Ferrers through to Maldon East, was handling over 1,000 wagons a year in the early 1920s. More intriguingly, among them were examples from the Highland, Great North of Scotland, Furness and Brecon & Merthyr systems.

NOT AMUSED

In an overly dramatic journalistic piece, a North Norfolk newspaper looked back at local railway history and to 1862 when the Prince of Wales acquired Sandringham House. According to the article, the Prince contacted John Valentine, the engineer of the Lynn & Hunstanton Railway and told him, 'We are, of course, very glad to have a station at Wolferton. But it is rather hard on my mother [Queen Victoria] as we haven't got a waiting room. She has to sit on her luggage on the platform.'

This is not an easy scene to visualise, but a waiting room was eventually provided to resolve the royal problem. And, as the Prince paid for it, he apparently felt entitled to use it as a lunch venue after a morning's shooting on the estate.

PANTO

A feature of the Railway Mania – when speculation in new railways rushed into madness – was that the nation's infant railway system even infiltrated the Victorian's insatiable appetite for pantomime. Three years before his eventual downfall, 'Railway King' George Hudson was satirized, along with railway speculation in general, in a pantomime entitled *Harlequin and the Steam King* put on at Sadler's Wells Theatre in 1846.

James Watt also achieved panto notoriety in the following year when a production depicted him devising pretty well the whole railway system from the inspiration of a steaming kettle while wondering how to win the hand of the blacksmith's daughter!

TIGER HUNT

Animals have frequently caused railway chaos, but none more dramatically than a tigress loose in Northamptonshire in 1877. The beast had been despatched from Broad Street in a low-sided wagon; unwisely as it turned out, for she escaped and promptly killed a couple of sheep to show her displeasure.

Hearing that the errant animal was prowling between Wolverton and Rugby, the Weedon stationmaster, in the best tradition, organised a hunting party of local gentry and garrison officers, unearthed an engine and set off in pursuit. The tigress was spotted and local people persuaded to act as beaters while the gunmen took potshots at the unfortunate creature peppering her with eight bullets and several charges of shot.

A BIT OF A SETBACK

Landowner obstruction, shortage of capital and construction obstacles were all commonplace problems experienced by early railway promoters. Some railways also encountered less predictable setbacks, among them the East Norfolk Railway (ENR).

Soon after the five main railway systems in East Anglia had combined in 1862 to form the Great Eastern Railway, the new company was actively encouraging new schemes for lines in the area north of its Ely–Norwich–Yarmouth route. One response was from the East Norfolk Railway, which secured an act in 1864 authorising a line from Norwich, through Wroxham and North Walsham to Cromer. Full of enthusiasm, construction work was begun in the following year.

As was often the case, the ENR contractor accepted shares as part of his remuneration but then died leaving his affairs less than tidy. The resultant legal hiatus prevented calls for instalments towards the line's £88,000 authorised capital, a mess which it took five years to sort out. Another seven were then to pass before the East Norfolk Railway finally reached Cromer.

MISSED!

As if the now defunct Fife Coalfield, with its risk of subsidence and resultant speed restrictions, was not enough, enginemen on the Aberdeen–Dundee–Cupar–Edinburgh route had to cope with Kinghorn Tunnel. The coalfield area had constrained their climb up from Thornton Junction to Dysart and instead of the free run down to Burntisland and on towards Edinburgh, trains had to reduce speed to 30mph through the tunnel.

The reason? The original tunnelling inwards from each end did not meet as precisely as had been planned with the result that the tunnel has a kink, a short reverse curve, near the middle of the tunnel. Ever since, trains have had to pass through at reduced speed.

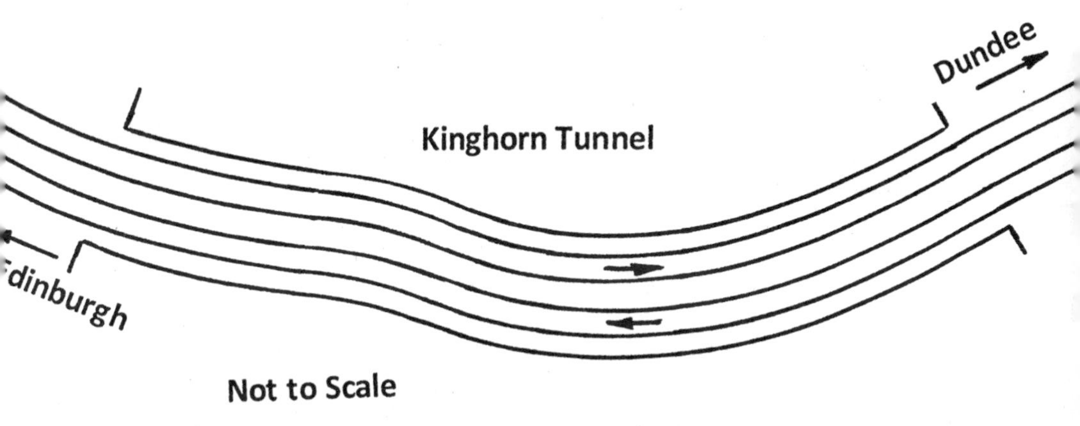

The profile of Kinghorn Tunnel. (Jim Dorward)

FLY SHUNTING

Fly shunting, in the form of repositioning rolling stock controlled only by its brakes, was largely confined to goods wagons. It was, however, authorised for coaching stock at several places where there was no more orthodox way for a locomotive to run round its train. One of these was Killin at the western end of Loch Tay, a modest place which had got its branch railway from the Callander & Oban line in 1886 thanks largely to the influence of the Marquis of Breadalbane. Jim Dorward describes the fly shunting process at Killin in the following example:

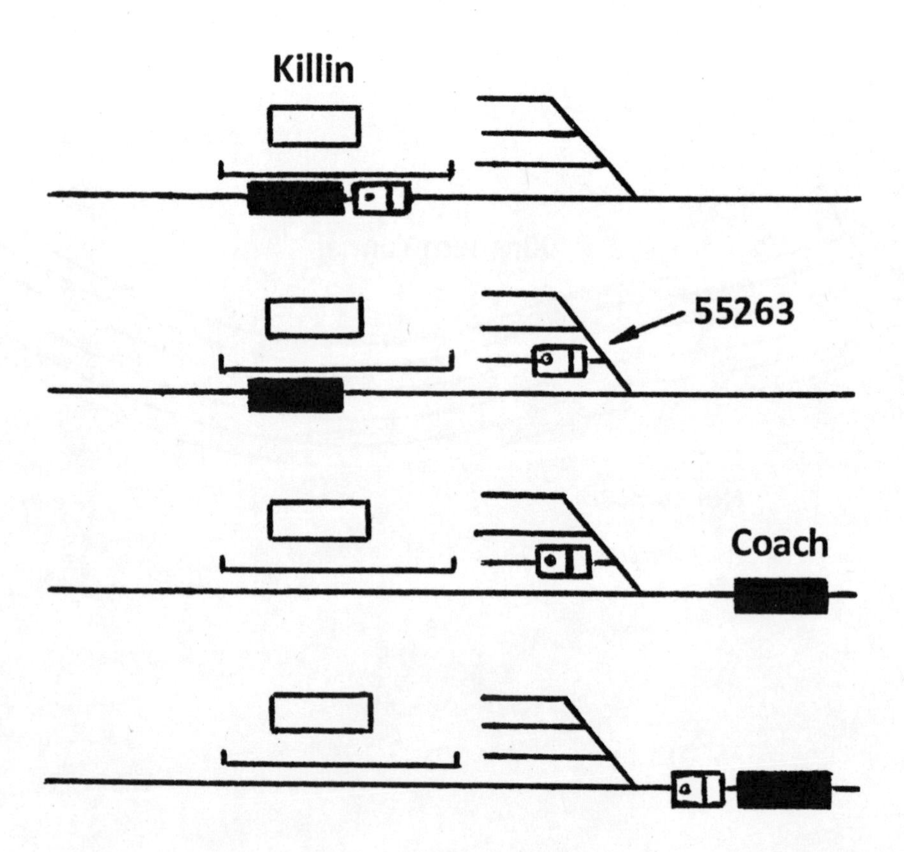

Diagram showing the fly shunting operation at Killin. (Jim Dorward)

It is Monday 1 August 1960. The Killin Junction to Killin branch train, composed of only one Thompson non-corridor second class coach with a guard's compartment, has arrived at Killin and its passengers have left the one-platform station.

The steam engine, ex-Caledonian 2P, 0-4-4 tank No.55263, has uncoupled, moved forward and then reversed into the small goods yard. With the points in the correct position, the guard rejoins the coach and releases the brakes. As the coach is on a slight gradient falling towards the points, it moves off, demonstrating the low coefficient of friction that exists between steel wheels and steel rails.

After the moving coach has passed the points by a slight distance, the guard applies the brakes and brings it to a stop. This allows the engine to leave the goods yard, couple onto the coach and move back to the station to wait for its next trip to Killin Junction, on the Oban line.

ANDERSON'S PIANO

The rugged Callander & Oban Railway had not long been opened throughout when trouble began to be experienced from loose boulders where it passed along the north side of the Pass of Brander on the bottom slope of Ben Cruachan. These large stones started moving on the mountain slopes and rolled down onto the line. Watchmen patrolled the railway but John Anderson, the indefatigable Secretary of the Callander & Oban Railway, suggested in August 1881 the erection of a fence-like wire screen. A breakage in the wires of the fence would then set at danger automatically the 'stone signals' at the entrance to the hazardous area. An experimental length was swiftly agreed, was tested in January 1882 and was then extended to the whole length of the line exposed to the stones, giving about 3¼ miles of wire screen fully operational from 17 April 1883. The sound of the wind in its wires gave rise to the nickname 'Anderson's Piano'. Drivers were told that so long as the wires were unbroken the signals would be 'off', but if the wires had been broken by a falling stone from the hillside, they would then be set automatically to danger, in which case a driver had to reduce speed and advance cautiously. A smaller section was installed between Callander and Strathyre.

Ironically, the southern section of the route, which was due for closure under the Beeching economies, was prematurely closed by a landslide at Glen Ogle. The normal signalling on the remaining route has now been achieved for a long time by Radio Electric Token Block (RETB), which has no traditional signals, but semaphore 'stone signals' remain in use. The need was underlined in 2010 by an accident with a rogue boulder, resulting in a carriage being left precariously suspended above the loch. More recently experiments have taken place with an electric sensor system to detect rock movements, which would benefit vulnerable roads as well as the railway.

A DOUBLE LIFE

One small railway halt, 6¾ miles west of Brighton, had two distinct lives serving two very different purposes. The first existence had its roots back in 1898 when a pioneer American filmmaker began his business on an area of the beach just south of the town of Shoreham-by-Sea. There, William Dickson built a studio and over the next decade was so successful that a need arose for local accommodation for the actors, technicians and others associated with the growing activity. Coincidentally, the London, Brighton & South Coast Railway had decided to relocate its carriage and wagon works from Brighton to a new home at Lancing and effected the move over the four years from 1908–12.

Despite the demise of the filmmaking activity, the demand for accommodation continued and the Lancing Carriage Works did quite well in disposing of its redundant railway carriages to eager buyers able to pay the £10 purchase price and another £3 to have the body delivered along the River Adur using lighters drawn by horses when the tide was low. The motley collection of these homes had continued to grow until it warranted its own railway station on the main coastline with the result that Bungalow Town Halt was opened there at the beginning of 1910.

Although the 'Bungalow Town' development continued as a holiday home location – not all that conveniently as there was no electricity and water had to be purchased at twopence a bucket – its train service was eventually withdrawn and the station closed on the first day of 1933, only for both to be reinstated eighteen months later, this time to cater for passengers using Shoreham Airport. The new lease of life as Shoreham Airport station lasted until a second closure early in the Second World War as the aerodrome changed its role from civilian to wartime use.

KIPPERS FOR BREAKFAST

In the fairly flat countryside of East Anglia, winter flooding is not all that unusual, despite the miles of drainage waterways. Summer flooding is a different matter and a problem rarely encountered in August. Nevertheless, in 1912, the last week of that month brought upon the railways of the Great Eastern system a bout of wild storms and subsequent massive flooding, especially in the area enclosed by the Waveney Valley line to the south and Wells-on-Sea in North Norfolk. Not that other places escaped, for trains had to be halted as far south as Melton on the East Suffolk line and Ramsey many miles to the west.

In addition to the immediate effect upon train running and the damage being caused to track and signalling, the need to suspend many train services stranded a large number of would-be passengers. Great Yarmouth was particularly badly hit with hundreds wanting to travel from Vauxhall and many not having enough money left after their holiday to find local accommodation while they waited for things to improve. They were allowed to sleep in carriages and the waiting rooms and fed there by the stationmaster and his staff.

Things were as bad at the other Yarmouth stations, with Beach badly affected by three blockages on the Midland & Great Northern route and some fifty people to be accommodated. Here again, every possible spare room on the station was made available and, somehow, three meals a day were provided for them. Not surprisingly in view of the location, fish featured prominently on the emergency menu including kippers for breakfast and fried fish for lunch.

Some passengers completed their journey despite the floods, but often by some unusual routes. A Liverpool Street–Cromer train, for example, only reached Norwich by traversing the Forncett–Wymondham line and then 'wading' through the swollen waters of a dozen streams heading for the River Yare either side of Hethersett. A service of cabs then took people on to Whitlingham for the final leg to Cromer. In fact, they were the lucky ones for the main line had to be closed subsequently because of flooding near Stowmarket.

One traveller on Monday 26 August set off from North Elmham to London, sat at Yaxham for two and half hours and was then carried back northwards for another long wait at Kings Lynn. An attempt to reach the Ipswich line via Bury St Edmunds ended in a return to Lynn and forward again via Newmarket and Cambridge, eventually reaching Liverpool Street at four in the morning after thirteen frustrating hours. Amazingly she got back to her home on the following day via Yarmouth, Stalham and a cab for the final leg.

400-TON SWING

On 8 July 1903, eight locomotives passed slowly over the 800ft, five-span railway swing bridge which had then been under construction on the approaches to Great Yarmouth for nearly four years. There must have been feelings of great relief all round as no significant deflection appeared, making the way clear for the new route linking Yarmouth with Lowestoft to be opened shortly afterwards. The line, double throughout except for the portion over the bridge, was an enterprise of the Norfolk & Suffolk Joint Committee, a rare example of co-operation between the Great Eastern Railway (GER) and the Midland & Great Northern, and one that gave the latter access to Lowestoft from its Yarmouth Beach station. The new 14½-mile coastal line also provided a connection into the GER's Yarmouth South Town and carried a lot of holidaymakers to and from the six intermediate stations and their holiday camps and other leisure venues.

For the final section into Yarmouth the North–South line had to cross the River Yare where it flowed out of Breydon Water and wound it its way through Yarmouth town to the sea. Hence the bridge, which was 800ft long, had five spans and had cost £38,453. At the southern end was the shorter swing section, which was powered by a small engine mounted high on the superstructure above and which could be opened to permit the passage of the numerous coasting vessels that then sailed to and from Norwich. Growing Broads holiday traffic in later years resulted in the bridge's dual 60ft passage ways being normally kept in the open position and then closed for any of the forty or so trains that once used the line.

The enabling legislation for the several earlier swing bridges in the area required them to be manned by the railway at all times and contained a proviso to the effect that if any vessel was delayed more than five minutes there was to be a fine of £5 per half hour. It was a yacht, the *White Seal*, that brought the Breydon Bridge its first problem, as a result of damage it sustained because the bridge was not open when the vessel got into difficulties. The railway was summoned by its London owner and fined £5.

Bad weather could turn Breydon Water into a fearsome place, but the bridge proved sound and even lucky for it escaped damage during the Second World War during one of the many bombing raids on Yarmouth. One German bomb exploded in the water beyond the bridge, but only after it had whistled through the girders without touching any of them. Sadly, the iconic Breydon Swing Bridge lasted only fifty years, being closed on 21 September 1953 and now replaced by a rather functional lifting-span road bridge.

EXCHANGE STATIONS

In contrast to the bustling activity of a station like Manchester Exchange, there were a few smaller places labelled 'exchange stations' which were very different in character. Mostly they served no local community and existed solely to facilitate the transfer of passengers between trains. At one time there were such exchange stations at Bala Junction where a change was necessary to reach Bala proper, at Barnwell near Cambridge and, in Scotland at Methven Junction between Perth and Crief and at Cairnie Junction on the Great North of Scotland line. Generally, tickets were not issued to or from these stations and they had no conventional access facilities.

Remarkable among these exchange stations was Roudham Junction, deep in the wooded area beyond Thetford on the line to Norwich. At one period Thetford had branches south to Bury St Edmunds and north to Watton and Swaffham and, in theory, passengers could change between the local services on these lines and the more important Ely–Norwich line. Not too many would want to do so with only one morning and two evening trains calling on the Swaffham branch. Roudham Junction was officially an exchange station only from 1902, and from 1920 main line services ceased to call except during the period of the coal strike the following year. Final closure came in 1932.

What the fate of Roudham Junction might have been if the original promoters' thoughts of a through route from Kings Lynn to Colchester via Thetford, Bury St Edmunds and Long Melford had ever been realised is just intriguing conjecture.

INTERLACED

Combining two lines of track in almost the space normally occupied by one is a rare, but not unknown, phenomenon. It has the advantage of avoiding the use of points and signals and certainly saves space, but does have its penalties in that the two routes so linked cannot be used at the same time.

An example of this situation occurred at the Glossop triangle off the Great Central main line between Sheffield and Manchester. There the two lines which approached the terminus from the different directions were interlaced for the last 400 yards permitting the signal box, along with the points and station junctions, to be better sited.

Another example existed on the East Coast Main Line at Walton, just north of Peterborough. There the tracks of the London & North Eastern Railway main line and those of the London, Midland and Scottish branch to Stamford were crossed by a fairly important road. Both routes had their own level crossing gates and the branch its own signal box but the width of the crossing was so substantial that the Up slow line was interlaced with a parallel freight line to avoid unwieldy gates or an unacceptable gap.

ADDING INSULT TO INJURY

In the flat, fertile countryside of the Fens there are, inevitably, a great number of occupation and accommodation level crossings where private roads or tracks cross the line or fields on either side are connected. They are not only the bane of the railway operator's life and not much loved by the users either, but also represent a high risk of conflict between fast-moving trains and local traffic, despite the various safety precautions.

On one very hot summer day, a diesel multiple unit, with all its windows open to let in a bit of air, ran into a trailer containing liquid manure as it crossed an occupation crossing behind a farm tractor. The speed of impact was around 15mph, despite the driver having made an emergency brake application, which threw some twenty or so passengers from their seats. No one was seriously hurt, just cuts and bruises and considerable shock.

In some ways, the worst part of the incident was that the tank of liquid manure was fractured and its contents flung through the open windows of the coach onto the passengers inside. Not a nice experience for them or the railway staff who all expressed their views pretty bluntly to the 'wretched' tractor driver.

NOSEDIVE

The Furness Railway route south-west from Ulverston towards Barrow crossed an area in which extensive iron ore mining took place. In 1892, near the Lindal iron ore sidings, a goods train was involved in a startling incident as it crossed over the high embankment there.

Driver Postlethwaite, on the footplate of an 0-6-0 goods locomotive, had just started some shunting work when his engine gave a sudden lurch. Looking down he could see huge cracks starting to appear in the ground below and he lost no time in shutting off steam and jumping from his perilous perch. No sooner had he done so than a large crater began to form and his engine started to slide chimney first into it.

Later in the day, efforts were made to haul the errant locomotive back from the pit into which it was now quite deeply embedded, but the hole just got wider and the task had to be abandoned. The engine just sank further and further into the old workings below the track until it finally came to rest some 200ft below the surface and had to be written off the railway books.

THE NAMES LIVED ON

The impact of the emerging railway network on stagecoach activity was huge. Competing long distance routes had to be abandoned, but the coach operators quickly found another role in providing feeder services for a time. A longer-term legacy was to be found on the LNER where former coach names were perpetuated on its steam railcars, most of which had a framed print of its coach predecessor in the passenger seating area. At one period stationmasters were authorised to pay out 5*s* to anyone providing historical information about the original road service.

The first railcar had been built in 1847 at Bow for the Eastern Counties Railway and was named *Express*. Railcars of various makes and designs became quite commonplace in the years after Grouping with five of the LNER's Sentinel-Cammell cars being named, stirringly, *Defence*, *Royal Sovereign*, *Tantivy*, *Old John Bull* and *Old Blue*. Another five Clayton cars followed the trend with four called *Comet*, *Chevy Chase*, *Railway* and *Transit*. The fifth was a bit of a let-down being called *Bang Up*. Locomotive engineer William Stroudley also forsook romantic names when he called his 1885 personal railcar *Inspector*.

EXTRAORDINARY SPECIALS

Once the early railways got the idea that there was money to be made from running special trains they went for it 'full regulator'. Three years after opening, the London & Brighton Railway (L&B) ran its first public excursion on Easter Monday 1844. The size of the train on leaving London Bridge had been increased at New Cross and Croydon and when it eventually reached Brighton, the original forty-five carriages and four engines had increased to fifty-seven vehicles and 6 engines!

A few months later there was another remarkable piece of L&B enterprise, this time not quite so much in the train arrangements as in the occasion. A flamboyant character called 'Captain Warner' had announced that he would demonstrate his 'invisible shell' off the coast at Brighton and several specials were run to convey government officials and spectators down to witness the event. In due course the vessel being towed in the demonstration certainly exploded and sank but there was a great deal of evidence that the event had been rigged. However, no doubt the crowd enjoyed the spectacle and the railway had collected a handsome bonus for its revenues.

MAN VS TRAIN

The original penetration of the Dartmoor vastness from the south dated back to an idea of 1818. This was the Plymouth & Dartmouth Railway, planned to bring granite down from quarries at King Tor and opened between there and Sutton Pool in 1823. It took 23 miles of track to find a workable route winding through the wild rising terrain and to cover what was just half that distance in crow-flight terms. The later 1883 GWR line, from Yelverton to Princetown, followed much the same route and even the best of the five weekday trains took forty-two minutes to complete the 10¼-mile climb.

In an example of the long tradition of racing against trains, a soldier and two friends alighted at Ingra Tor Halt one day in November 1955 and set off to cover the uphill mile to King Tor before the train got there.

The pioneer Plymouth & Dartmoor Railway had to tackle this tough Dartmoor terrain before it could be opened in 1823.

Despite the engine driver taking up the gauntlet and cutting several minutes off the eleven allowed for the 2¾-mile journey, the three challengers won the contest and were able to rejoin the train and complete their journey to Princetown, no doubt well pleased with their achievement.

JELLICOE SPECIALS

Beginning in February 1917, a fourteen-coach special train began a daily service over the 717 miles between Euston and Thurso. It was a journey which took a wearying twenty-one and half hours. The trains, which continued to operate until 1919, came to be known as Jellicoe Specials as their purpose was to carry naval personnel and stores to Admiral Jellicoe's grand battle fleet at Scapa Flow.

The huge Scapa Flow base was only one of several served by the Highland Railway, a system that was placed under great strain in the First World War, not helped by the fact that less than a fifth of its 272 miles between Perth and Thurso was double track. The whole area north and west of Inverness was considered military territory and passengers were not even allowed station access without a special permit.

FIRE

The steam locomotive had an inherent risk of causing fires. Line-side fires were frequent and, indeed, continue to this day. Nowhere was the fire risk greater than on the lines which ran into dock areas in the era of sailing ships, built and equipped as they were largely with combustible materials.

The Millwall Extension Railway was opened across the Isle of Dogs by the London & Blackwall Railway in 1871–72. Designed to serve the new West India and Millwall docks and facilitate the development of North Greenwich, its short run south from Millwall Junction was in four sections, 5 chains at one end and 31 at the other belonging to the London & Blackwall, with a 41-chain section and 52-chain section in the middle owned by the respective dock companies. Because of the fire risk to timber yards and vessels in dock, the dock companies would not permit steam operation and so the first portion of the line was operated for the first eight years with a tram-like coach pulled by two horses, with a steam engine then completing the journey. When locomotives had to go back to the main line for repairs the fire had to be drawn and the engine hauled 'dead' over the dock companies' lines, again by horses. The Blackwall Company had also to take fire precautions on its main line near Regent's Canal Dock where it was required to provide 'a light iron roof' over the track.

When steam locomotives were accepted over the whole Millwall Extension route in 1880, they were, for a time, unusual in carrying enamel plates advertising Pears Soap. Another curiosity came when rail motors were introduced, one of the three purchased from the GWR proving to be abnormally long and encountering serious difficulty in negotiating the curves on the line.

"THE 4-31 IS RIGHT BEHIND YOU SO DON'T STOP TO TALK TO ANYONE!"
HANDS OFF ROAD TRANSPORT
VN

AVOIDING ACTION

In 1861, the year before the railway line from Kings Lynn to Hunstanton was opened, its engineer J.S. Valentine decided to take his family with him on a trip from their lodgings in Kings Lynn to the small coastal resort at the end of the line. Father, mother and six children duly took up their places on makeshift seats on a large contractor's trolley with a couple of workmen to pump its handles. They had only covered half of the 15-mile journey when a works train was spotted approaching fast on the single line. The low trolley would not have been easy to see and so some smart action was necessary to unload the passengers and luggage and then dismantle the trolley to lift it out of the way in time.

ALL PART OF THE SERVICE

As the British holiday habit grew, tourist traffic became increasingly important to the railway companies, especially to the smaller concerns such as the Talyllyn Railway. One facility it provided was a service of wagonettes to take visitors on from the Abergynolwyn terminus to view Talyllyn Lake. From there they might take another to reach the Corris Railway, use it to get to Machynlleth and there take a Cambrian Railways train back to Towyn.

Rather more unconventional was the provision of a slate wagon, which was hauled up to Abergynolwyn by a timetabled train and then available for passengers to coast down on as desired after the return of the last train.

PITCHED BATTLE

In the early years of railway promotion and competition, rivalries frequently became quite intense. None more so than what has been styled 'The Battle of Saxby', referring to the Midland Railway's Syston & Peterborough line scheme and the opposition it encountered in trying to survey a route through Lord Harborough's Stapleford Park. In the process, the opposing sides – railway surveyors and estate staff – came into violent physical confrontation over four days from 13–16 November 1844. Excerpts from the newspaper coverage of the events give some idea of the ferocity of the encounters which took place.

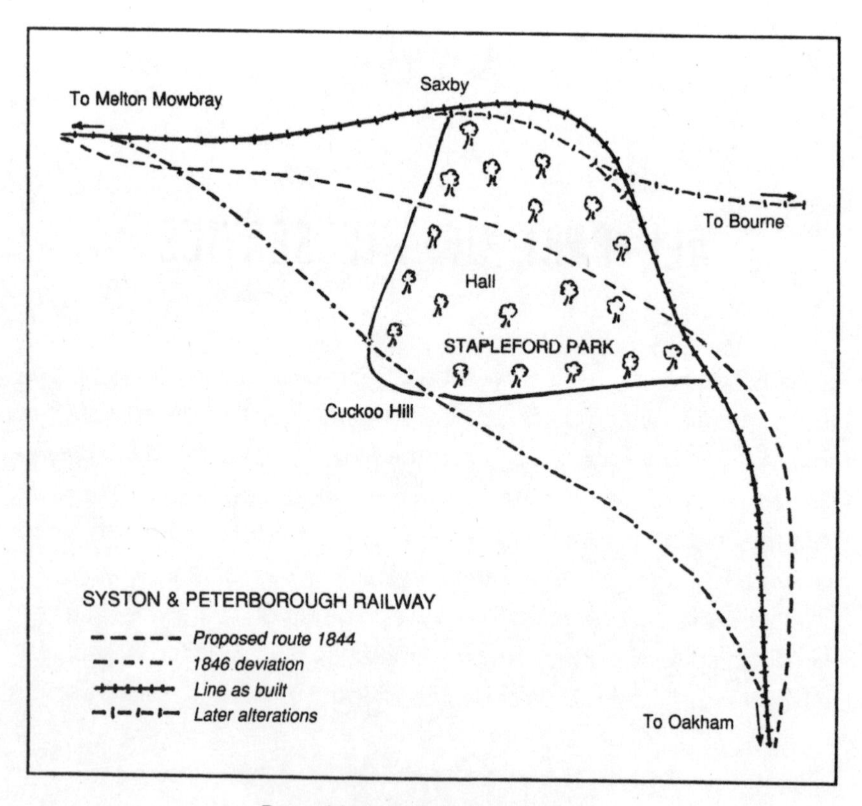

Routes of the Syston & Peterborough Railway.

An account relating to Thursday 14 November dealt with events after a small skirmish had occurred on the previous day and a large contingent of the earl's men had 'taken prisoner' the modest survey group. This time:

> Half the defending forces wedged themselves on the Melton side representing a formidable living barrier. The engineer and officers of the other party of men drew up with their backs to the forces of the earl's party; the rear ranks to rush upon their own friends and drive them through the hostile array. Men's bodies were seen from the pressure to spring high in the air over the heads of the parties. Mud bedraggled the clothes of all. A breach was made in the line of defenders and the chain carried through in triumph, but was immediately seized hold of and broken.

The Friday was quiet, but brought news of another assault planned for daybreak the next day. The railway party turned out to be several hundred men this time, gathered from as far afield as Peterborough. The estate officials had called up their reinforcements from the surrounding villages but, not knowing where the blow would strike, the defenders were scattered. Even so, they had brought up five fire engines to douse the intruders and some small cannons from Lord Harborough's yacht to intimidate them. That day's account includes the peak of the engagement when:

> Brown, the lock keeper of the Oakham Canal, a powerful man, rendered great service to his lordship, sending his opponents head over heels at every blow; the noise was so great that it was heard in villages two miles away. The spikes of the railway party were thrust into the sides of the defendants of the park and after a battle of about five minutes, and many broken heads, wounded faces and sides, the lower grade of the intruders gave way.

After this, the focus of events moved to litigation. Eventually the railway did get its line but had to seek a Deviation Act and go round Stapleford Park and then alter that again after forty-four years to ease the severe curve it entailed.

UNDER THE INFLUENCE

When the Midland Railway finally got fed up with the delays its trains encountered in having to reach London via the Great Northern's metals between Hitchin and Kings Cross, it decided upon its own independent access to the capital. Initially, the company would hardly have expected the design of the new terminus to be largely determined by the dimensions of barrels of beer brewed miles away in the Midlands. Nevertheless that was the case.

To clear the Regent's Canal the approach line to St Pancras had to rise 17ft above the Euston Road. With Burton beer in mind, W.H. Barlow abandoned the original plan for a solid lower foundation for the new station in favour of creating a street level goods warehouse with the platform area above. Access between the two would be by hydraulic lift.

The support columns in the warehouse were placed 29ft 4in apart in order to accommodate the maximum number of barrels of beer, which arrived on a central track and were lowered by lift to the storage and delivery area. This valuable freight business lasted for nearly a hundred years.

PATERNALISM

Railways were built in an age when paternalism was strong. Their promoters included many a country gentleman accustomed to obligations towards employees and tenants and many religious men concerned with the wellbeing of their flocks or just of their souls. A sense of duty was strong, not only in the railwaymen and their families, but also in the obligations felt by those who required them often to work in isolated locations.

Not only was there a railwaymen's co-operative shop on the island platform at Riccarton Junction at one period, but the North Eastern Railway also operated 'church trains' which provided free passage to Sunday services at Hawick one Sunday and Newcastleton the next in order to allow railway folk to worship. Non-railwaymen could use these trains at half fare. The North British Railway, too, made concessions for its staff travelling to attend church services and at Hawes, in the Yorkshire Dales, services were held in a platform waiting room.

Beyond Okehampton: the approach to Meldon Junction was heralded by this impressive viaduct over the Okement River. The historic 120ft high, 1874 structure is scheduled as an ancient monument.

Trains were used to supply water to quite a few remote railway outposts, particularly in Scotland, while at both Meldon Junction, near Okehampton and Sugarloaf Summit on the meandering Central Wales Line stops were scheduled to allow railway wives to go shopping.

On the West Highland Line children living in isolated homesteads were picked up and taken to school at Rannoch or Fort William but when the numbers grew even further the Highland Railway set up its own school in an old carriage. On the same route there were also special provisions for water supplies and for arranging a locomotive to convey a doctor in the event of an emergency.

TUNNEL VISION

Bores, one might think, would be straightforward and simply boring, but some were certainly not. There was even a view in the early railway days that no line of significance should be without one. The directors of the pioneer Canterbury & Whitstable Railway certainly felt this way and accordingly rejected one fairly level route for their line in favour of one which required a tunnel through Tyler Hill. The end result was that the gradients limited the use of locomotives and necessitated two rope-work inclines. Until more powerful locomotives became available, delays were, unsurprisingly, frequent. Still, this was in 1830 and was therefore Britain's first conventional railway tunnel.

Two years later, the Leicester & Swannington Railway opened and handed later railway operators the problem of the very narrow Glenfield Tunnel. Its profile meant that the line could only be traversed by bespoke carriages fitted with inward opening doors and window bars to prevent anyone leaning out. Gates at either end of Glenfield Tunnel were closed at night. So, too, were those at Tyler Hill, which had the further odd characteristic of having differing sections because the three building contractors employed had differing ideas.

Restricted tunnel clearances occurred both on main routes like the Hastings line and Worcester to Hereford and on minor ones. Using a former tramway tunnel seemed a good way of reducing construction costs but, on the Burry Port & Gwendraeth Valley line in South Wales, it added one more need to engineer cut-down locomotive cabs in its later years.

And Tyler Hill was by no means the only unnecessary tunnel. Audley End was one built to placate a landowner while the 1,100-yard tunnel on the approach to Newmarket was conceded to answer the fears that trains would frighten the horses and kill racing on the famous racecourse there. At Coulsdon, the London, Brighton & South Coast tunnel was the perceived answer to the avowed risk of frightening the inmates of the nearby lunatic asylum with the 'fearful and terrifying noises' the trains would create.

SLEEPERS

The first proper sleeping carriages were introduced by the North British Railway on its Glasgow to Kings Cross service, in 1873. They contained two compartments each with three curious armchairs, which could be contorted into an upside down profile for sleeping.

Before this, guards on special trains like the Irish Mail hired out a primitive package of two rods and a cushion which could be stretched between the seats for lying down, but that and a rug hire service were about the best overnight travellers might expect prior to the introduction of sleepers.

HIGHLAND RIOT

Protection of the Sabbath was apparent nowhere more than in the Highlands of Scotland. The Dingwall & Skye Railway was to encounter this with its courageous route through the wild and beautiful area to the Western Isles at Loch Carron. And so, too, was Queen Victoria who was refused fresh horses at the Achnasheen station hotel during a Sunday visit to Loch Maree.

The western end of the line from Dingwall had initially to be truncated at Strome Ferry, short of its Kyle of Lochalsh destination, owing to lack of funds. That meant the railway having to provide its own steamer service for the arriving Skye sheep. Large quantities of fish were also landed there, but things went badly wrong when a Highland Railway special train was provided to carry them forward on a Sunday. Incensed Calvinists promptly lay in rows upon the exit line of rails and would not move until their protest had been recognised and alternative accommodation arranged.

BRUNEL'S WRATH

Brunel worked himself hard and expected as much from others. When this was not forthcoming he could be pretty forthright in expressing his dissatisfaction, as one of his assistants realised when he received a note saying:

> Plain, gentlemanly language seems to have no effect on you. You are a cursed, lazy, inattentive, apathetic vagabond, and if you continue to neglect my instructions, and to show such infernal laziness, I shall send you about your business.

THE LONG AND THE SHORT AND THE TALL

Not surprisingly, Britain's longest distance between stations was in Scotland, between Tulloch and Rannoch on the West Highland line. These were 17 miles 24 chains apart, although a couple of private stations intervened. From Rannoch to Bridge of Orchy was another 15 miles 48 chains, again excluding private stations.

At the other end of the distance scale, it was only 30 chains from Aberdeen North to Schoolhill on the Great North of Scotland Railway and not much more over the river bridge linking Culrain and Invershin stations on the Highland.

Excluding the Snowdon Mountain Railway, the highest point on a conventional railway was again in Scotland, namely 1,484ft above sea level at Drumachter Pass. Somewhat surprisingly, second ranking, just 111ft lower, fell to Princetown on Dartmoor.

RUNWAY CLEAR

The interlocking of signals and points has long been a prime railway safety feature. One of its more unusual applications was in Northern Ireland where the railway passed near a wartime airfield at Ballykelly, east of Loch Foyle.

As the war progressed, bigger bombers came into service and the runway at Ballykelly had to be lengthened to handle the four-engined Lockheed Liberator aircraft based there. This involved extending the concrete surface over the railway track with just narrow slits left for the train wheels to run in.

Avoiding the risk of collision between trains and planes was done by setting up a special signal box with a link to the aerodrome flying control. The 'Runway Clear' signal could only be sent over this when the railway approach signals were locked at danger.

FARE DODGERS

The contest between railways and those passengers intent on not paying a proper fare has gone on since railways began. Initially ticket collection methods were not uniform, the Stockton & Darlington Railway (S&D) for example collecting them at the time of admission to the train.

For a short time the S&D had a problem with people paying for a short distance but intent on going further. To counter this, the railway put them all in the same coach and uncoupled it when it had covered the distance they had paid for.

A less simple situation confronted the Great Northern Railway on one occasion when female conjoined twins arrived at Kings Cross from Edinburgh and proffered a single rail ticket. A lengthy debate ensued but the matter was eventually dropped as being too complex to resolve.

BEDS AND BAILIFFS

Beginning with the financial crises of the 1840s, many railways ran into money problems. The failure to respond to share calls, high land costs, construction overspends and disappointing earnings all featured in the causes. Stringent economies were one consequence and another was a frantic haste by creditors to distrain upon railway assets.

A typical sufferer was the East Anglian Railway (EAR). Cutting costs was tried, including the provision of beds at stations for porters to save the cost of night watchmen, but it was not enough. One contractor, insisting on security for his account, had twenty carriages and a quantity of materials deposited in a shed at Kings Lynn as his surety. Around the same time several creditors took possession of such of the EAR property as they could put their hands on resulting in a court battle between them and the bond holders over which group had the prior claim. A major factor in this small railway's troubles was the clear intention of its Eastern Counties Railway neighbour to swallow it up, which it eventually did.

An especially bizarre example of penury problems occurred in 1870 when, in Ireland, a County Clare court order put bailiffs on the footplate of one train's locomotive. They were ejected just a few days later by the men of the County Galway sheriff as a result of yet another court order.

'PURSUED BY THE FORT'

This curious phrase and its novel setting appeared in a pamphlet published in 1862 by one William Bridges Adams. Adams was an engineer, writer and inventor who started working life in the coach-building industry, spent some time in South America and later turned to railways, contributing designs for springs, axles and even steam rail motors.

Apparently inspired by the role of the Federal gunboats in the American Civil War, Adams used the pamphlet to advocate not only defensive rail routes around London, but also coastal defence railways equipped with rail-mounted gun forts supported by wagons for back-up infantry. This, in fact, anticipated the role of the rail-mounted gun in twentieth-century conflicts, but suggesting an invading army might be 'pursued by the fort' was, perhaps, a concept too far.

DUAL PURPOSE

Dual-purpose bridges carrying rails and road are not all that rare. The notable example over the Tyne at Newcastle is still in use and others existed over the New Cut at Bristol and the Nene at Sutton Bridge. However, one dual-purpose structure, which stands out from the others is the 670ft long viaduct crossing 90ft above the Luxulyan Valley in Cornwall. The ten-arch edifice, now a scheduled ancient monument, was completed in 1844 for Squire Joseph Treffry as part of his early exploitation of the area's china clay riches. Treffry's enterprise marked the beginning of a vast china clay rail network and the viaduct was a central feature carrying not only his tramway line but also a water leat to feed no less than thirteen waterwheels.

The Treffry beginning formed the basis of the Cornwall Mineral Railways system and embraced a new harbour at Par with canal, tramway and incline links inland. Essential to the inland rise of the route was the Carmears tramway incline which was itself powered by a 34ft waterwheel turned by the water carried over the Luxulyan Viaduct.

HEAVY RESPONSIBILITIES

Being employed by one of the early railway companies was much to be desired, but was no sinecure. Operating staff on the Eastern Counties Railway, for example, had to understand, sign for and abide by no less than 108 regulations. Among these, was No. 77 which read:

> Every engine is to be supplied with the following articles and the engineman is held responsible for the care of the same; namely, a complete set of screw keys, a large and a small monkey wrench, three cold chisels, two hand hammers, a pinch-bar, a screw-jack, a towing chain, two coupling-chains with hooks, a signal-lamp, a set of signal flags, a gauge-glass lamp, a large and small oil can, a bucket, two pump-clacks, a gauge-glass tube, six tube plugs, a coke-shovel and set of fire-tools, and a quantity of flax spun yarn, waste and twine.

EASTERN COUNTIES RAILWAY.

SIGNALS

AND

REGULATIONS,

20TH DECEMBER, 1846.

Books of a previous date are incorrect.

EASTERN COUNTIES RAILWAY.

RULES AND REGULATIONS.

Every Officer and Engineman, before he shall be allowed to serve on the Line, shall sign these Regulations, and for disobedience to which, he will be punishable as for an offence against his employers and against the law.

SIGNALS.

HAND SIGNALS.

DAY.

1. The Signal *All Right* is shown by extending the Arm horizontally, so as to be distinctly seen by the Engine Driver.

ROAD AND RAIL

The dream of a vehicle capable of use on both rail and road and with simple transfer between the profiles of the two has surfaced several times. The 1960s, for example, saw the brief appearance of a large transferable freight container, labelled the Road Railer and exhibited at a number of major railway depots before lapsing into obscurity.

With railways experiencing increasing road competition, the dual-mode concept received a lot of attention in the 1930s as part of the general search for a more effective small capacity unit than the conventional train of locomotive and coaches. In 1931, the London, Midland & Scottish Railway (LMS) produced its Ro-Railer, designed in-house and based on an ordinary single-decker bus body with a 6ft 3½-in road track outside conventional rail wheels. After a road run the vehicle would be run up

The Michelin pneumatic-tyred railcar *Coventry Pneumatic* in rail mode, with a party of officials at an unknown location.

on a special ramp and, once the rail wheels rested on the track, the road wheels were slid clear vertically and locked in position. The changeover between rail and road or vice versa could be accomplished in two and a half minutes, and the vehicle could attain speeds of 60mph on the road and 70mph on rail.

The Ro-Railer was tried out in service between Harpenden and Hemel Hempstead and between Blisworth and the Welcombe Hotel at Stratford-on-Avon, which the LMS had just bought for £70,000. Passengers travelling from London could transfer to the new dual-role vehicle at Blisworth and at Stratford it would then use the special ramp to transfer to normal roads for the journey on pneumatic tyres to the hotel. This courageous new concept was, however, not completely successful and the idea was dropped, the same fate overtaking a similar London & North Eastern Railway experiment with goods vehicles on the West Highland line in 1934.

After the Ro-Railer, the LMS did not abandon its interest in pneumatic tyres for small passenger units and, utilising French experience, introduced such a Michelin vehicle in 1935. Known as *Coventry Pneumatic* it could reach 65mph, but did look quite curious in having its driving controls housed in sort of conning tower sticking up 2ft 6in above the roofline. This idea, too, came to naught.

PRICEY

The railway shareholders' manuals of the 1840s cast some interesting sidelights on railway activities and ambitions. Under the chairmanship of George Hudson, the so-called 'Railway King', the Eastern Counties Railway (ECR) was clearly aiming high, with Tuck's 1847 shareholders' manual declaring, 'The Company is in treaty for the Ambergate, Nottingham

& Boston Railway, which if concluded will enable them to occupy all Lincolnshire and the adjacent districts.' It was to be many more years before the ECR's successor even penetrated Lincolnshire.

The entry went on to record the company's introduction of the electric telegraph on the Cambridge and Colchester lines where 'upwards of 180 miles of wires have been placed, and 60 instruments fixed at the various stations.' Extolling the virtues of the system it was said that 'on a recent occasion a celebrated London physician was in communication with a Norwich physician and through the agency of the electric wires, actually prescribing for a patient whose life was in danger.'

The consultation would have cost the London physician 7*s* 6*d*, the same as a second-class fare to either Cambridge or Colchester.

THE OLD WORSE AND WORSE

This was the apt nickname for the Oxford, Worcester & Wolverhampton Railway (OW&W) built to link West Midlands manufacturers with their London markets. Its route between Worcester and Oxford entailed a 6-mile climb, mostly at 1 in 100 and through the 887-yard Campden Tunnel, a section that produced a truly bloody battle during the construction period between Brunel's men and a failing contractor. Brunel was involved because the Great Western was backing the line and, probably, because the contractor R.M. Marchant was a relative and former assistant.

By July 1851, Brunel had had enough of the resident contractors Williams & Marchant and dismissed them in favour of Peto & Betts. Marchant refused to withdraw his workmen so Peto & Betts then collected a large force to take over the works with a dramatic result, which *The Times* reported in the following words:

Early on 21 July 1851, this peaceful stretch on the western approaches to Campden Tunnel would have been crowded with navvies heading towards confrontation with the men of a sacked contractor.

> On reaching the Worcester end of the tunnel Mr Cowdery with his gang of 200 men from Evesham and Wyre was met by Mr Marchant who dared anyone of Peto & Betts men to pass the bridge on pain of being shot, Mr Marchant himself being well supplied with pistols. Mr Cowdery, exercising great forebearance at the unseemly conduct of Mr Marchant, told his men on no account whatsoever to strike a blow. Mr Cowdery, finding that all expostulation was useless and Mr Brunel giving peremptory orders for Peto & Betts men to proceed and take everything in line, a rush was made to the men which after a few seconds was repelled with great force by Marchant and his men, the consequence was that several heads were broken and three men had their shoulder dislocated. Up to this time the navvies had not called in to requisition the picks or pickaxes or shovels, but a man in the employ of Marchant having drawn pistols he was seized upon and his skull near severed in two.

The account goes on to record a stalemate with policemen, soldiers of the Gloucester Artillery and magistrates arriving on the scene. There was a further brief melee even after the reading of the Riot Act and more GWR reinforcements had turned up, and in a final skirmish one man had 'his little finger bitten off' and another 'his head severely wounded'.

Arbitration eventually settled this tumultuous affair and work could proceed, but a later OW&W flirtation with the London & North Western Railway's alternative link from Oxford to London was to cause the Great Western further, albeit less violent, frustration with its rebellious prodigy.

HIJACK

The Great Northern Railway (GNR) penetrated right into Midland Railway territory by using an agreement with a smaller company with a long title: the Ambergate, Nottingham & Boston & Eastern Junction Railway. However, the GNR made the mistake of using one of its own locomotives for an afternoon departure from Nottingham on the day the service began, whereupon, as the *Nottingham Journal* put it, 'the Midland Company's engineers placed several locomotives before and behind it, and took the trespasser prisoner'. The action was then successfully defended in court by the Midland Railway because it had only agreed for the use of their station by the Ambergate company.

Until the matter got sorted out the stranded Great Northern locomotive remained in captivity.

PUSH AND SHOVE

In the power struggle between the individual railways around Manchester, the East Lancashire Railway (ELR) had a major opponent in the Lancashire & Yorkshire Railway (L&Y). The rivalry brought the two into a bizarre confrontation at the beginning of March 1849. It took place at Clifton Junction where the ELR line from Accrington and Bury joined the L&Y line from Bolton and used it for the final run into Manchester. When the East Lancashire company changed the arrangements by which the L&Y could check how many of its passengers used the final section the latter took what might be labelled decisive action.

On Monday 12 March, the L&Y placed a huge timber baulk on the running line on the Manchester side of the junction and reinforced this blockade with a train of empty coaches, supposedly to take on passengers forced to disembark from the ELR train from Bury, which the obstacle had compelled to halt. A contemporary newspaper report describes what happened next:

> The first act of the East Lancashire Company's servants was to remove the baulk of timber, and this they did without hindrance. The next proceeding of the East Lancashire party was to run forward their train, and attempt to force before them the Lancashire & Yorkshire blockading train. The others, however, having put on their breaks (*sic*) and brought another engine up in the front of their train, which they had detached from an express from Manchester that had come up while the previous proceedings were going on, were able to maintain their ground. The East Lancashire Company by this time having brought up a heavy train laden with stone, on their other line, now caused it to run forward, and take up a position on the shunt across the Lancashire and Yorkshire's up line to Manchester, exactly abreast of the blockading train on the other line of rails. Thus the Lancashire & Yorkshire Company's double line of rails was completely blocked up.

The two rival railways then compounded the position by using further trains, as they arrived, until there were eight trains blocking the two lines for ½ mile. The whole scene was observed by policemen but no violence occurred and after a couple of hours the L&Y local man decided enough was enough and broke the deadlock by removing the obstacles on his side so that the delayed trains could proceed and the matter be resolved at higher levels.

BLOCKADE

In 1850, two small railway schemes intent on linking lines from Bolton and Clitheroe came into headlong conflict with the East Lancashire Railway at Blackburn. At issue was the need of the Bolton, Blackburn, Clitheroe & West Yorkshire Railway, later the Blackburn Railway, to pass its trains through the East Lancashire company's Blackburn station. The latter demanded an exorbitant toll based on a rate of 6 miles for the ¾ mile involved and wider rivalries and ambitions aggravated the basic conflict of interest between the two parties concerned.

A dramatic consequence of this situation was described in the *Blackburn Standard* on 26 June 1850 in the following terms:

> When the first train from Clitheroe arrived on Saturday morning it was found that a complete blockade of the points had been effected at the junction with the East Lancashire Railway near Turners Mill at Daisyfield and that upwards of 200 navvies had been brought on the ground by the East Lancashire Company with several engines and a heavy train of stone wagons to enforce an obstruction on the public – whose convenience, by the way, the East Lancashire Company have always professed to be peculiarly careful. This disgraceful blockade continued during Saturday, only one train being permitted to pass on payment of the very extravagant toll, but even that toll was subsequently refused and the notice was given to the Bolton Company that their engine would be removed from the line in case they attempted to pass again over the East Lancashire Company's Railway.

The level crossing, signal box and disused platform at Daisyfield, Blackburn where passengers affected by the 1850 'blockade' would have had to alight.

It seems that the two railways had been arguing about the toll level for months with both prevaricating and then losing patience, bringing them to their extreme position. The Bolton management had reached the point of threatening to force a way through the controversial section and violence had become a real possibility. The sufferers were, as usual, the poor passengers who, for a time, had to change trains by a walk between the two routes.

Behind the scenes lurked the East Lancashire's great rival, the Lancashire & Yorkshire Railway whose managing director Captain Laws now warned that 'a game had been commenced at which two could play.' This dire threat was enough to diffuse the situation with the Bolton company reluctantly agreeing to pay the toll demanded on the understanding that it would eventually be adjusted to the ruling the Railway Commissioners would be asked to make on the subject. As it turned out, the toll fixed by them was based on a 2-mile rate rather than the 6-mile rate originally demanded.

ACCESS

There were hundreds of 'railway hotels', from the basic to the opulent and catering for an equally varied clientele, from weary commercial travellers to the rich on holiday. The one thing they had in common was ease of access; the commercial traveller not wanting to carry his samples further than necessary and the rich finding it unthinkable that they should have to carry luggage or walk too far. Hence examples like the Great Western Hotel at Paddington and the Royal Station Hotel at Hull Paragon where platform and hotel foyer were virtually contiguous. At Saltburn, the Zetland Hotel was opened in 1863 when the line from Middlesbrough to Redcar was extended to the coast. The frontage provided magnificent views over the sea aided by a central turret topped by a telescope room while at the rear the railway platform was extended right to the building's back wall and provided with a covered exit walkway.

The former Railway Hotel at Nailsworth was typical of its kind.

At Saltburn the railway-owned Zetland Hotel had its own platform access.

The Royal Hotel at Bideford was linked directly to the Southern Railway station platform.

When the Great North of Scotland Railway opened the Cruden Bay Hotel in 1899, it was ¾ mile from the railway's branch line from Ellon so the company built a 3ft 6½-in electric tramway and two special tramcars to link the two. The cars could carry coal and laundry as well as passengers and continued to do so for nearly twenty years after the railway lost its passenger service in 1932.

Motor vehicle links between stations and hotels were commonplace – a survival from the stagecoach days. One was still operating in 1945 between the station and the Cross Keys Hotel in St Neots proper, some 2 miles away.

OTT

There was great excitement among the senior figures in the Great Western hierarchy on Thursday 31 May 1838. After a host of construction trials and setbacks, the directors had assembled at the temporary Paddington terminus for a ceremonial trial run to the station east of the Thames at Maidenhead. With Daniel Gooch's *North Star* in charge, the journey was successfully accomplished at an average speed of around 28mph. This excellent start was then marked by a cold but substantial luncheon to which some 300 people sat down in marquees and toasted this prelude to the public opening scheduled for the following Monday.

So carried away by the occasion was Bristol director T.R. Guppy that on the return journey he climbed up onto the roof of the leading carriage and proceeded to make a precarious and flamboyant demonstration of his own by passing the length of the train in this dramatic fashion. Literally, going over the top!

ROCKET WARNING

Colonel William Congreve developed a form of rocket artillery, which was introduced into the British Army in 1805. The early versions were, to say the least, pretty erratic in performance but were later improved and certainly had a major shock effect deriving from their noise, speed and fiery trajectory. This weapon was sufficiently notorious for it to be featured in one pundit's warning about the dangers of the emerging railways. One writer declared:

> Can anything be more palpably ridiculous than the prospect held out of locomotives travelling twice as fast as stage coaches. We should as soon expect the people of Woolwich to suffer themselves to be fired off upon one of Congreve's Rockets as trust themselves to the mercy of such a machine going at such a rate. We will back Old Father Thames against the Greenwich Railway for any sum.
>
> We think that Parliament will in all Railways it may sanction limit the speed to eight miles an hour which is as great as can be ventured on with safety. As to those persons who speculate on making railways generally throughout the kingdom … we deem them and their visionary schemes unworthy of notice.

The reference to Old Father Thames is no doubt an allusion to the growing number of paddle steamers carrying passengers to downriver destinations as far afield as Margate. The first such vessel had arrived in 1814 and by 1820 the number of boats being operated had exceeded twenty.

A CASE OF CONFUSION

At one time, travellers liked to showcase their wanderings by adding to the labels on their luggage. Many of these labels were highly decorative, but the practice was not appreciated by transport operators, as the *Daily Mail* reported in June 1914 under the heading 'Luggage Label Fetish'.

According to the report the London & North Western Railway had warned against the confusion which might arise from this practice, especially when luggage was sent off in advance of a journey. It was revealed that porters, faced with a number of conflicting labels, generally chose the cleanest as being the most likely to be the most recent. That this did not always prove to be the case was apparent in an example quoted in which the kitbag of a passenger travelling from London to a suburban destination found it had been sent to the West Coast of Africa!

THE PRIZE-FIGHTING GAME

The early railway companies often seemed more interested in profit than in probity or scruple. In the areas around London, the practice of running special trains for prize-fighting contests was quite commonplace, however much some might decry the practice.

In theory, bare knuckle fighting was banned in 1826, but the practice remained popular, albeit harder to stage. In this situation the railway companies were prevailed upon to hire special trains to convey spectators to venues that had to remain secret to all but the aficionados of the so-called sport. On the day appointed, the special might be obliged to hasten on to an alternative destination if the authorities turned up at the first one.

Then, crossing into another county was a well-tried device for evading the local police force. It was to be 1868 before legislation finally ended this particular railway activity which presupposes a level of locomotive, coach and staff availability which borders on the incredible.

Essex was one favoured area for these fighting events, especially the remote landscape of the Thames Haven branch where intruders could easily be spotted. Kent provided further examples with the South Eastern Railway carrying over 2,000 people to a contest east of Headcorn in 1859 and then another to Strood two years later. When the London, Chatham & Dover Railway ran a similar special to Meopham, police intervention meant a move on to Sittingbourne to complete the fight programme. Clergymen, police and even the Home Secretary protested to the railways about their support for such a dubious activity but the revenue available proved a stronger argument.

NAMING

At first railways followed the stagecoach practice by giving their locomotives stirring names. The first two delivered to the infant GWR, by canal to West Drayton in November 1837, were called *Premier* and *Vulcan*. Class names then followed with the 1839 *Morning Star* leading to more 'star' varieties. Another direction was taken with the machines ordered from Fenton, Murray & Jackson of Leeds, which first appeared in 1840 and in which *Charon* was later joined by *Hegate, Lethe, Phlegethon, Ganymede* and others. What the enginemen thought of this is not known.

Unsurprisingly, it was not long before nicknames emerged, with some quite expressive ones. One rather appropriate example was *Mac's Mangle* for J.E. McConnell's London & North Western Railway single of 1849; a name derived from the damage its wide outside cylinders and frames did to some station platforms. McConnell's 1851–61 express engines were treated no more kindly, being called *Big Bloomers*, an allusion to the dress revolution prompted by Mrs Bloomer's new ideas for pantaloons, which were perceived by footplatemen to be slightly too revealing. The slightly smaller machines which followed from 1854 were inevitably known as *Little Bloomers*. Other notable people to be 'honoured' in this way included the 1880s Jersey-born actress Lillie Langtry who gave her name to Great Central 4-4-2 locomotives known as *Jersey Lilies*.

Once under way there was no stopping the nickname habit which not only gave informal names to locomotive types (*Fat Nannies* – Lancashire & Yorkshire 0-6-0 saddle tanks), but also to whole railways (*The Old Worse & Worse* – the Oxford, Worcester & Wolverhampton Railway), trains (*The Honeymooner* – the 1.30 p.m. Paddington to Penzance), places (*Khyber Pass* – a cutting at Wood Green), equipment (*Newcastle Hook* – a shunting pole) and practices (*Peg Up* – place the signalling block instrument in the 'Train on Line' position).

BEACHED

At one time or another, trains have been halted by floods, snowdrifts, storms, subsidence, fallen trees, straying animals, metal thefts and even leaves on the track. A stoppage of a different sort occurred on the newly opened Hoylake Railway on 29 September 1866.

The Hoylake Railway opened as a single line from the outskirts of Birkenhead along the north-west coastal plain of the Wirral Peninsular to Hoylake on 18 June 1866. The objective of its promoters was to open up the low-lying, marshy terrain along the shore of Liverpool Bay to residential development, something achieved more fully under later Wirral Railway ownership. Traffic was sparse at first and what the fledgling concern really did not need was to suffer a minor sandstorm blowing over its exposed line. Large quantities of windblown sand completely bogged down the locomotive due to haul the 5.15 p.m. train from Hoylake on 29 September. With no spare power available there was no alternative but to cancel it and to suffer the ignominy of having to hire horses to haul the next service two hours later.

EULOGY

Originally conceived as a means of restoring prosperity to the small Hertfordshire town of Buntingford, the secondary ambition of the Ware, Hadham & Buntingford Railway, that of extending to reach Cambridge, was never realised. Instead, the 13¾-mile railway, opened northwards from the Hertford East branch to Buntingford, was destined to fulfil the life of a typical rural branch line. An unusual difference from even the most ambitious of railway schemes was the lengthy exuberance of a local poet in the 140 lines penned to mark the enterprise. Aired by the *Herts Guardian* it marked a financial rescue which led to the opening, with Great Eastern Railway operation, in July 1863:

It was the prime of summer-time
When our new railway was begun,
And o'er the line with ray benign
Shone forth the glorious July sun.

Hard by the hill that skirts Westmill
Hundreds that day did congregate,
With banners gay, in proud array
And all becoming pomp and state.

From far and near through Hertfordshire
The living tide of people pour'd,
And many there had come from Ware
From Hadham and from Buntingford.

To the same spot from Hall and cot,
They came of every rank and class
Knight of the shire, and country squire
And lady fair and rustic lass.

Thus met, we raised a song of praise
And on the air so soft and calm,
To heaven upborn that summer morn,
Arose the grand Old Hundredth Psalm.

Then kneeling round upon the ground,
We prayed to God our work to bless;
'O Lord look down, and deign to crown
Our honest labours with success.'

Next off the land, by maiden's hand
A sod was cut with silver spade,
Hard by the hill that skirts Westmill,
And in a tiny barrow laid.

Then from the place with female grace
She lightly on the planking trod,
Conveyed the load along the road,
And forthwith wheeled away the sod.

Her task complete, she sought her seat;
We shouted all with one accord,
As we stood there 'Three cheers for Ware,
For Hadham and for Buntingford.'

Just at that time rang out the chime
Of village bells, when, as before,
The exulting crowd with voices loud
Cheered thrice and then gave one cheer more.

Thence to the tent the party went
(Which had been spread in case of wet);
And as was meet, here, the elite
Had 'dejeuner a la fourchette'.

Grace duly said, speeches were made,
And many a loyal toast proposed
And some did stay long after day
Had passed and evening round them clos'd.

Nay, 'tis averred that sounds were heard
Strange echoes on the midnight air;
'Come pass the wine along the line
Of Hadham, Buntingford and Ware'.

Before we part, join every heart
In flowing bumpers round the board;
Hip hip horray our new railway
For Hadham, Ware and Buntingford.

But never yet on foot was set,
Or in the country or the town,
A noble plan for good of man
But some would always cry it down.

And one or two, I know not who,
Misunderstanding our design,
Foreboding spoke with raven croak,
Against the opening of the line.

Predicted they 'it would not pay'.
  But said 'it was no fault of theirs'
'And as the rail was sure to fail
  They must decline to purchase shares.'

But three long years mid hope and fears
  Still persevered our Railway Board,
To make with care the line from Ware
  To Hadham and to Buntingford.

Ne'er did shirk their arduous work,
  Nor toil, nor money, did they spare:
And now my friend, they've gained the end,
  At Hadham, Buntingford and Ware

Ages ago – '*Finis Coro –*
  Nat opus' said a learned sage,
And time has tried and verified
  That just remark in every age.

The line is done which they begun,
  And carried out the distance through,
Their task is o'er, and what is more
  The line shall safely carry you.

The other day in grand display
  Cover'd with glory and renown,
They yoked their team, put on the steam
  And made the journey up and down.

Their iron steed, when at full speed
  Swift as the winged lightning flew;
The glittering cars, they shone like stars,
  Adorned with flags, red, white and blue.

Just like as when, that citizen
Of ancient credit and renown
A wondrous wight, John Gilpin hight
Did ride this way from London town.

I do record, our Railway Board
Did ride a faster race than he,
Through Hadham, Ware and Buntingford
And many more went out to see.

And as you know three years ago
When first we did commence the line
We had that day a dejeuner,
When 'twas finished, we did dine.

The Railway Board at Buntingford,
A sumptuous banquet did provide,
And viands fine, and choicest wine
They to the company supplied.

The cloth removed – the most approved
Toasts they proposed, and healths they drank,
And faces glowed as champagne flowed,
And no man from his bottle shrank.

Not that I mean it was a scene
Of rude convivial excess,
But wine gave birth to noisy mirth,
And somewhat boisterous happiness.

Still over all that festival
One railway spirit did preside,
And his mild sway they did obey
As friend, director and as guide.

Nought there could be but harmony
And hearts that beat with one accord,
Love filled the chair that day for Ware,
For Hadham and for Buntingford.

But time would fail to tell the tale,
And wearily the muse would lag on,
Of all that passed from first to last
That day within the George and Dragon.

But take the hint and read in print,
What the *Herts Guardian* has to say
In language fine about our line,
And patronise our new railway.

Invest your cash and fear no crash
And let the company record
How fast and well the shares will sell,
For Hadham, Ware and Buntingford

And never fail to go by rail,
(There's nothing to alarm the fare),
On our railway is no delay
At Hadham, Buntingford or Ware.

The opening junketing over, the Buntingford line settled down to a mostly routine existence. It was involved in the filming of *Happy Ever After* in 1953 and *Girls in Arms* three years later and then had to cope with some severe winter weather in February 1958 when Braughing stationmaster H.E. Ribbons recorded:

> Snow drifts resulted in the 5 p.m. train from Buntingford, with about 50 passengers aboard, being snowed up between Standon and Hadham. We had a light engine here and it was decided to use it to try to push the train through to Hadham. We managed to get just beyond Standon when we ran into deep

drifts and could not move in either direction. I was carrying the train staff and started to walk the 3 miles to Hadham.

A gale was blowing, it was dark and I could not keep my handlamp alight, the snow was up to my waist in places, and my Wellington boots were full of melted snow. I found the disabled train, which by now was almost covered in snow. I advised the passengers and crew the position and continued my exhausting walk, sometimes on the permanent way, and sometimes on the fields. I was jolly glad to arrive at Hadham and receive a good stiff brandy from the SM Mr Brian. Guard C. Hills from Buntingford had walked from the disabled train to Hadham. He was so exhausted that he had to have medical attention.

PASSENGERS ON TOP

Railway opening day ceremonies were designed to impress, to recognise the efforts of the directors and engineers, reward the shareholders' confidence and excite the interest of potential users. The most privileged of these groups enjoyed special treatment and their accommodation on the inaugural train was usually quite extravagant. Others often did not do as well.

The Stockton & Darlington Railway held its formal opening event on Tuesday 27 September 1825 and began the proceedings with a demonstration of loaded wagons being hauled up and down the Etherley and Brusselton inclines at the western end of the line. Then followed the special train journey to Stockton with a large train marshalled behind *Locomotion* with George Stephenson acting as the driver. The fourteenth of the forty vehicles was a special coach for the directors and senior shareholders but all the others were wagons, one with the engineers and surveyors, fourteen with workmen, six 'filled with strangers' and the rest loaded with either coal or flour and carrying 'passengers on top'.

At Darlington twenty wagons were detached, permitting the workmen to indulge in 'victuals and ale' and some coal to go for distribution to the poor. Some of the 'passengers on top' had also had enough but there were plenty anxious to replace them. The workmen's wagons filled up again with people, two others took on the Yarm Town Band and the train then moved on again to Stockton where its occupants were greeted by cannon fire. The directors and important guests marched to the town hall where they were provided with a lavish meal and then enjoyed (or endured) no less than twenty-three toasts.

INCLINE SKILLS

Railway traffic operation demanded a degree of skill frequently unrecognised in the level of pay for the job involved. Nowhere was this more the case than on the numerous rope-worked inclines which once linked mines, pits and quarries with the main line rails at a lower level. Such lines were highly individualistic, with working methods subject to no external controls and based on the peculiar needs of the location and the skills, or otherwise, of its engineer.

On a typical incline a raft of perhaps four loaded descending wagons would be used as the power to haul up empty ones by linking the two by means of a rope, which passed around a drum at the top of the incline. Some basic form of signal gave the 'right away' for releasing the brake on the drum and the two wagon 'trains' would then gather pace, passing one another by means of a short stretch of double track halfway up the slope. By the time the loaded wagons reached the bottom they could be travelling at as much as 15mph and the shunter there had to accurately and firmly strike the pin that released them from the rope. Other shunters then used their sprags and scotches to bring the loaded wagons to a halt in a very physical chase where a slip could bring serious consequences.

Any failure by the lower shunter would be catastrophic. The descending wagons would hurtle on into those standing in the sidings, shattering everything in the resultant collision. The ascending ones, no longer retarded by the release of the rope below and the level of the top section of track would tear into the drum housing with equally devastating consequences.

There were, of course, many variations on this hair-raising operation. On an incline near Ludlow a shunter came down on the leading loaded truck and pulled on a wire to release the pin with the wagon then being slowed by a short section of track with an upward slope.

The Cromford & High Peak Railway (C&HPR), opened in two halves, in 1830 and 1832 to exploit the minerals of the Derbyshire uplands, had no less than nine inclines, eight worked by stationary engine and one by a horse. An accident, in 1888, on one of these inclines was quite spectacular. A coupling failure resulted in a brake van hurtling down the last incline at the Cromford end of the line and, failing to negotiate the curved approach to the canal wharf there, it flew through the air in a dramatic trajectory that cleared the canal and the main line railway beyond and ended with a crash into the line-side field.

Later C&HPR had a sequence of gongs installed on the incline slope. These would be struck by an arm on the descending wagons so that if the tempo exceeded the normal rhythm, the shunter at the canal approach points could divert the runaways into a crash pit.

CAB COMFORT

The extension of the railway system into the remote area north of Inverness owed much to the owners of great estates who were anxious to improve access to their properties. Prominent among these was the Duke of Sutherland whose home was the thirteenth-century Dunrobin Castle on the North Sea coast of Sutherland, about halfway between Inverness and John O'Groats.

The Duke had had a hand in achieving the opening of the line northwards to Golspie in 1868 and then funded its forward extension from there, initially to a private station for the castle. For a short time this section was operated as the Duke's own private railway with its own 2-4-0 tank engine which the nobleman drove himself on occasion.

Working was soon taken over by the Highland Railway but the Duke retained some extraordinary special privileges including the right to operate his small private train over the main line so that he could arrange for special guests to be conveyed between Inverness and the castle in style. The most privileged might expect to ride in the cab of the second ducal locomotive, an 0-4-4 tank obtained in 1895 from Sharp Stewart, named *Dunrobin* like its predecessor, but fitted with a nicely upholstered seat in an elevated position at the rear of the cab.

Until the motorcar took over, liaison with the Highland Railway would produce an operating path for the working of the Duke's special when guests were due to arrive by train at Inverness or by steamer at the Kyle of Lochalsh. The estate's engine driver would prepare *Dunrobin* and collect the coach from the station and duly set off to make the pick-up in style. During their stay the special might also make a local journey conveying members of the shooting party to a more distant part of the estate.

The original private station was rebuilt in 1902 in a simple style and then continued to feature in the main line timetable, 86 miles from Inverness and with a request service of five trains a day. Happily *Dunrobin* and a special coach have survived into preservation.

UNEASY BEDFELLOWS

The North London Railway was one of quite a few early schemes that saw advantage in a long and comprehensive title. It started life as the East & West India Dock & Birmingham Junction Railway. Reaching the London docks in 1852, an unusual feature was that, in return for a sum of £10,000 per year, it permitted the Northumberland & Durham Coal Company to operate its own engines and wagons over the line for the purpose of carrying seaborne coal brought down by ship from the north-eastern coal staithes. Two operators on one railway was bound to lead to priority clashes and the coal company's rights had to be bought out in 1859, a move which cost the North London Railway £43,000 but gave them five more locomotives of various types.

A PENNY A YEAR

With an early journey from London to Bristol taking anything up to nine and a half hours there was a clear need for a refreshment stop en route. To avoid adding to the massive costs already incurred in building the railway, the Great Western Railway leased the provision of refreshment accommodation at a midway point, Swindon, for just one penny a year. In due course, a quite palatial pair of buildings was provided by the chosen contractor, one on each side of the station's four through lines and connected by a footbridge. The lease required the GWR to stop all but express trains and specials 'for a reasonable period of about ten minutes' so that passengers could alight and make their purchases and the lessees redeem their original outlay.

As early as 1942, the directors of the railway were already regretting the deal because of a welter of complaints about the new facilities, and Brunel, never one to shun a blunt opinion, wrote a much quoted and very sarcastic letter suggesting that he had, indeed, never been given 'inferior coffee' but was surprised that the lessees bought such 'badly roasted corn'.

Over the years, the Swindon refreshment rooms varied from excellent to poor and twice the railway went to law to seek relief from the irksome delay burden with which it had saddled itself. Quite apart from the obvious drawback of having to add an extra ten minutes to the timings of so many trains routed via Swindon, the sheer herding, cajoling and urging needed to get passengers separated from their sustenance and back into their carriages was a permanent nightmare.

ON THE MOVE

To tap the agricultural potential of its rural East Anglia homeland, the Great Eastern Railway opened a number of minor lines in the late eighteenth century. One of these was the Wisbech & Upwell Tramway constructed along the route of the earlier Wisbech Canal and brought into use as far as Outwell in 1883, with an extension to Upwell in the following year.

Running for some 6 miles southeast from Wisbech, this was primarily a freight line – coal inwards and fruit and vegetables outwards – but a passenger service operated until the end of 1927. Initially six trains a day, this unusual railway used tramway-type coaches with end steps and gangways, hauled by small tank engines fitted with side panels, cow catchers and warning bells.

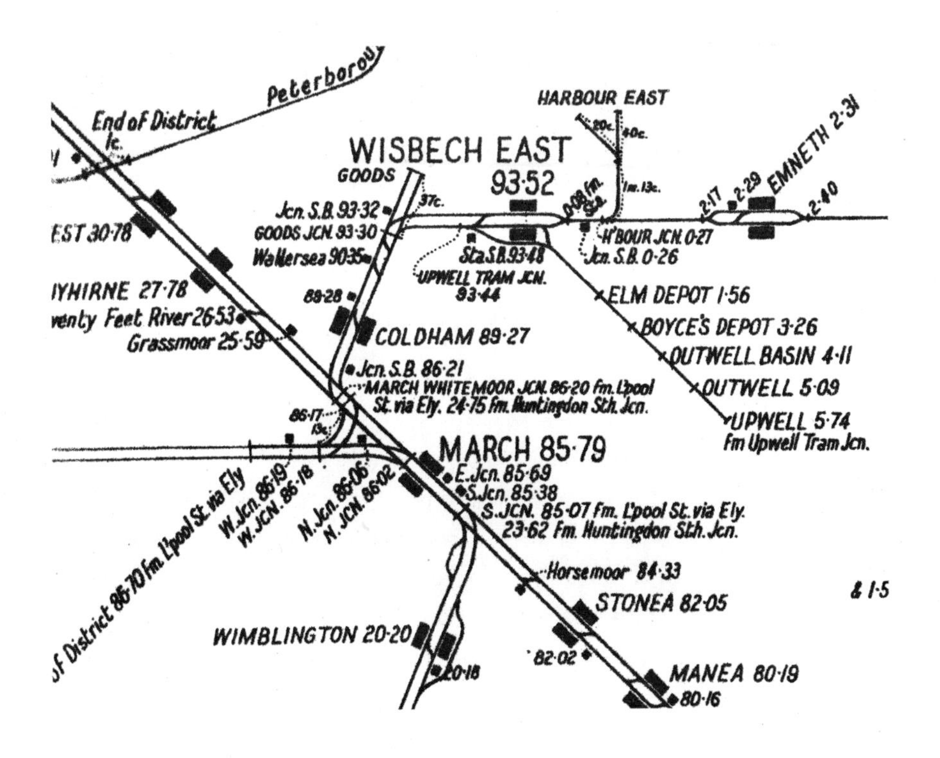

The freight service continued strongly until the 1950s but then gradually shrank as road competition increased, eventually ceasing in 1966. Even in the later years, however, the little line dealt with huge quantities of produce during the Cambridgeshire fruit season, moving the loaded vans back to Wisbech in order to be worked forward to Whitemoor marshalling yard to connect with the evening express freight trains to London and provincial centres.

The aim of the local growers was to continue the day's crop harvesting as late as possible and still secure early arrival at the markets to achieve the best prices. No time could be wasted on the journey so the branch railway provided an office van for the Upwell trains with goods clerks setting out from Wisbech in the morning, with one dropped off at each station so that the waybill preparation could be done as consignment notes were collected from the arriving loads. The process continued at a frantic pace even as the returning trains headed back to Wisbech so that none of the clerical processes should result in delays.

The use of a mobile office was quite unusual but hectic document preparation was a feature of many of the railways' urgent traffics and may even have spelled the end of the copper plate writing that was the norm in earlier years.

CONFUSION

The rich, sun-blessed land south of the Mendip Hills is ideal for growing strawberries, a fact that led to the railway from Yatton to Wells becoming known as The Strawberry Line – a name it still carries in its later incarnation as a popular footpath. In its heyday this GWR branch, which carried a modest through-passenger service over its 31½-mile route from Yatton to Wells and then on to Shepton Mallet and Witham, loaded many wagons of Cheddar strawberries in the fruit season.

Amid a small development of retirement accommodation, Sandford Station has been superbly restored to function as a small museum-cum-heritage centre.

The Yatton–Wells section was opened by the Bristol & Exeter Railway in 1869 and some of its fine buildings survive, both station and goods shed at Cheddar and at Sandford where a fine station heritage centre has been created. Winscombe, the next station on from Sandford, was originally known as Woodborough and a well-founded story there recounts that the station nameboard was originally erected upside down.

More confusion existed at Wells, the country's smallest city and one which, nevertheless, boasted three stations, all within a very short distance of one another. This was the result of the meeting of three different railways and a confusion of routes, connections and two different gauges, which took five years to sort out.

CIRCULAR ROUTES

London's Inner Circle underground line is by no means the only roundabout route to have existed. It also had its companion Middle Circle (Aldgate–Mansion House) and Outer Circle (Broad Street–Mansion House), while the Southern Railway ran both Waterloo to Waterloo trains and others out from London Bridge and back again. A circular double was created by the SR trains running London Bridge–Norwood–Selhurst, with those from London Bridge to Crystal Palace inside. The London & North Eastern Railway (LNER) Fairlop loop was another roundabout example.

Away from London, the LNER electric trains provided a basic ten-minute service out from Newcastle Central to Tynemouth via Benton and back via Wallsend, taking just under an hour for the all-stations run over the 20½ miles. On Glasgow's Cathcart roundabout route the modest morning and evening trains needed around thirty minutes for just 8 miles.

Roundabout services, not always with through trains, existed around Nottingham and with plenty of time to waste one could travel out from and back to Bristol Temple Meads via Pilning Low Level and Severn Beach. On the other side of the country a leisurely circle could be travelled to and from Norwich Thorpe by way of Coltishall and County School – a route described locally as 'Round the World'.

TAKEN BY SURPRISE

On 5 April 1931 the 42,348-ton liner *Empress of Britain* was to sail down the Clyde on her way to her new owners, the Canadian Pacific Railway. She had been built in John Brown's Clydebank yard at a cost of £3 million and now, with five tugs to help until she was clear of the widening Firth of Clyde, was headed out on her maiden voyage.

A good opportunity, thought the London, Midland & Scottish Railway (LMS) district office, to run a couple of special trains for Glasgow people out as far as Langbank on the opposite side of the Clyde to Dumbarton Castle. The distance out from Glasgow was only 15 miles and Langbank itself, although a good viewing point, was no more than a tiny village at that time. So it was decided that two trains would be ample, one to leave at noon and the second one some fifteen minutes later. Rolling stock was duly arranged along with a couple of 0-6-0 goods locomotives to provide the motive power.

In the event the crowd that began to arrive at 11 a.m. and form queues at the booking office windows at Glasgow Central got bigger and bigger until it numbered some 11,000 people and hurried arrangements had to be made to find the coaches, motive power and staff to run more trains. Altogether twelve specials set off for Langbank. A certain amount of delay and confusion occurred in making the arrangements and then along the line, especially at Paisley where only the normal rostered Sunday staff were on duty. In the end every passenger was accommodated and although the journey took forty minutes everyone was able to marvel at the sight of the majestic 760ft-long white vessel leaving her birthplace.

Used for her short life on the route between Southampton and Quebec with a cruising programme during the winter, the *Empress of Britain* eventually fell victim to a German U-boat in October 1940.

HONEYMOON COMPARTMENTS

Nigel Gresley, before he took over from the redoubtable Henry Ivatt as chief mechanical engineer of the Great Northern Railway in 1911, was Ivatt's carriage and wagon superintendent at Doncaster Works. This experience was to stand him in good stead when the LNER came into being as a result of the 1923 Grouping and inherited a motley collection of passenger carriages from the various constituents that were brought together by that legislation.

In the year following the Grouping, the new railway announced a massive investment in new rolling stock and the first fruits of this intention came into public prominence with the introduction of the new *Flying Scotsman* train in October 1924. In addition to the new-look triple restaurant car set with all-electric cooking equipment, the first-class accommodation included some semi-private coupés, which were soon being called 'honeymoon compartments'.

More carriage improvements followed in the next few years including some groundbreaking innovations, prompted in part by the introduction of an all-summer, non-stop service running between Kings Cross and Edinburgh in 1928. Understanding that passengers would need extra facilities when confined to their train for eight hours, interior furnishings received the attention of a specialist – a newsboy offered newspapers and magazines on board and a hairdressing salon duly followed. Even more novelties appeared in 1932, perhaps prompted by the standards of the great ocean liners of the period. A small cocktail bar was provided, some sleeping cars, which included a shower compartment, appeared and full width windows became more commonplace on the prestige trains. Nor were third-class passengers neglected for they got folding armrests so that the old four-a-side seating could be eased to three when demand was not at its peak.

An especial novelty was the addition to the two early afternoon departures from Kings Cross and from Edinburgh of equipment to offer passengers recorded music or BBC broadcasts. Using a rotary converter coupled to the train's 24-volt lighting system and a 30ft aerial along the roof of the guard's compartment, a wireless receiver housed in a shockproof box was able to wire radio programmes from Daventry or Brookmans Park to the dining saloons and adjoining compartments. Stewards there offered headphones, which were presented in a sealed bag to ensure hygiene and then sent for special disinfection after each run. A modest charge of 1*s* was made for the hire.

TICKETS

By the standards of the time, the early railways were quite significant enterprises. Once built, they had to find recruits for a wide variety of tasks and while some were available from the ranks of displaced coaching activities, the shortfall had to be made up wherever staff could be found. Inevitably some were not fully literate.

To make sure that staff with difficulty in reading and writing could deal with the tickets they might have to handle, some tickets had colour on the reverse to indicate the destination station. Before the growing network enabled passengers to exchange between lines, a lone railway might show the image of a fleece on the ticket to show it related to a journey on the up line and something like a cotton bale for one on the down.

ROYAL STATION

When local interests prompted an extension of the railway network from Kings Lynn to the small coastal development at Hunstanton on the edge of The Wash, they were not to know that one of its stations was to be regularly used by royalty and distinguished guests. Just a few months before the new line opened in 1862, the then Prince of Wales purchased a house and estate at Sandringham that was to become a favourite retreat for the royal family. From this event arose a tradition of royal trains to and from Liverpool Street, all subject to special train notices and a whole paraphernalia of spotless engines and stock, standby locomotives, extra staff and a whole bevy of senior operating officials to ensure that no delay or other hitch occurred on the royal journeys.

The new 'royal station' at Wolferton led a schizophrenic existence, catering for its normal local service and the extra summer holiday trains plus becoming the focus of a great deal of official attention when a royal train was to run.

The early saga of the royal waiting room [recounted on p. 116] was followed by the provision of a brand new station on 1898. In a black and white Tudor style, its royal waiting rooms were furnished in a lavish but conservative manner. Over the years they were to be used by notables of every kind, from royalty and heads of state to politicans and the like. Visitors to Sandringham house parties were frequent at one period. There were sad occasions and also some bizarre ones like the incident of the circus elephant which uprooted the lamp post to which it had been tethered and had a few moments of wild freedom before trotting calmly back to its cage.

Another item of local railway folklore avers that Rasputin visited the station while Russia's Tsarina Alexandra was staying there as a royal guest in 1909. The strange monk had travelled across Europe, journeying across the Channel by packet steamer; spending some time in London and then heading north intent on seeing his patron. Unimpressed by this odd character, about whom they knew nothing, the Wolferton staff refused to believe that he would be welcome at Sandringham House and sent him on his way.

Another Wolferton incident occurred in 1936 after the death of King George V. To convey his casket from Sandringham House to Wolferton station for the waiting funeral train, a gun carriage was hurriedly smartened up with a new coat of paint. Unfortunately the paint was not given time to dry properly so that when the time came to lift the casket off it was found to be well and truly stuck. Quick thinking saved the day by having the gun carriage moved out of sight and the casket forcibly prised off, the whole operation proceeding so well that only five minutes' delay were incurred.

DICKENSIAN

Well into the nationalisation era, a typical railway goods office would have much in common with the sort of counting house portrayed in a Dickens drama. A continuous high desk with a series of lift up lids was likely to parallel the wall contours, each segment with a high stool in front of it, all of which were without a back except, possibly, those of the chief inwards clerk and his chief outwards counterpart. Their stools might well have a low back as a mark of their status while many of the others might be so worn by generations of railway bottoms as to make sitting on them more akin to perching and represented a constant danger of sliding off.

The chief clerk, probably of Class III grade or higher depending on the size of the depot, would enjoy a simple, conventional chair and a table with a couple of drawers for the tools of the trade. These would include a round ruler, pen and ink with several spare nibs and a bevy of copying pencils. Using a red one would certainly upset the auditors who claimed sole use as one of their prerogatives. By virtue of his supreme authority, the chief clerk would sit near the coal fire, which was the only form of heating.

The whole activity was, of course, paper intensive. Inevitably this produced a demand for somewhere to store the consignment notes, invoices, delivery sheets and countless other forms after they had been acted upon and many a goods shed lock-up was filled with these wrapped in dusty parcels. At one time the former boardroom of the original Bristol Committee of the Great Western Railway was totally full of old claims section papers, each with its attendant quota of dust and grime.

What a boon the advent of carbon paper represented. Before that, copies of accounts and other documents were made in large bound books with lined front pages for indexing and a host of tissue sheets onto which documents had to be copied by a process of dampening, followed by the use of an oiled sheet and of a heavy press for squeezing the image onto the next available page in the book. Too much moisture and the ink ran, too little and the copy was poor. This paper and press process was still in active use at Great Malvern as late as 1975!

The stationery cupboard, kept locked of course, and with the key carefully guarded by the chief clerk, was an Aladdin's Cave of items, from tiny 'omnibus' forms with a number of printed questions, instructions or information on them to huge returns for balancing the monthly accounts. There were envelopes with multi-address panels, large bottles of ink, rubbers and even the wartime economy issue of narrow metal tubes for accommodating the stub end of a pencil when there was not enough of it left for fingers to grip. Elsewhere in the office, well worn by constant use, would be large books of instructions and regulations like the *Classification of Merchandise* which indicated the rate to be applied to every conceivable commodity and the *Handbook of Stations* which recorded every station and its facilities.

The clerical procedures were complex and clumsy like the process of preparing abstracts from the invoices and designed for use by the Railway Clearing House in dividing the revenue earned by a consignment between the companies which had participated in its journey. Almost everything was done by the writing process; GPO telephones were rare and their costs so carefully watched that only the chief clerk was likely to be allowed to use one. Any attempt to use the railways' own 'omnibus' circuits was usually a frustrating and often abortive procedure.

Not that the typical small to medium booking office was any better. It, too, had its forms along with instruction books like the *Passenger Classification*, which recorded how to charge a 'prisoner travelling with escort' or a 'groom accompanying a racehorse'. Printed, blank card, paper, cloakroom etc., tickets had to be stored, 'face value' items kept secure and cash carefully locked in the safe after each balance until banked or sealed in a cash bag for sending off on a specified train. Many booking offices also dealt with parcels traffic which required a pot of turgid glue kept warmed for pasting on parcels stamps, luggage labels and the like.

Facilities? Very few, just a kettle perhaps. Unless one counts being able to take a key when heading over to the platform lavatories and thus saving the penny coin needed for access. Privileges? A few, including 'Privs', or more properly privilege ticket vouchers, which could be presented for reduced rate travel, and a few free passes annually. And, if a passenger could be persuaded to pay a shilling for protection by the Railway Passenger Assurance Company, a penny accrued to the selling clerk to supplement his modest salary – not to be sneezed at!

Truth be told, the most comfortable place on many a station was the porters' room and being allowed to slip in occasionally was a much sought after privilege.

RAILWAY TIME

For up to forty years, railways operated to times that frequently differed from those of the places they served which worked to local times based on solar observation for that particular place. Greenwich Mean Time (GMT) did not become universal throughout the country until 1880, but as early as 1840 the Great Western Railway had realised that it had to have a stable time basis for compiling and operating to train timetables. Other railways followed suit and, after a Railway Clearing House decision in 1847, all trains ran on London Time, or Railway Time as it became known.

A ten o'clock time synchronisation telegraph signal ensured all station clocks followed suit and guards set their watches before joining the first train of their roster.

It took the provinces quite a while to fall into line. At Bristol, for example, the local time was eleven minutes later than GMT and trains could easily be missed if the intending passenger was consulting a watch set to local time. The time duality continued uneasily, but gradually many local clocks began to display both times and the 1850s opening of telegraph offices, which received an hourly GMT signal, hastened the end of the time confusion.

SPECIAL TRAINS

One of the delights of earlier railway days was the sheer variety of trains on display. Not just the Pullman and other express passenger services but 'locals' pottering around the system and countless excursions, often with a motley collection of coaches garnered from wherever they could be found. There were postal trains, pigeon specials and freight workings ranging from Class C fully-fitted strings of wildly swaying meat or fish vans to a line of milk tanks, of incredibly long rakes of coal empties or, sometimes, just a few wagons pottering about behind the local 'pick-up' service. Add in the civil engineer's contribution, breakdown trains, the odd 'officers' special' and the Chapman weed-killing trains and the observer need never have a dull moment or be short of a scene for his camera.

From time to time something exceptional might appear such as an out-of-gauge load or a snow-clearing train with the plough leading and heading for action. Major agricultural shows brought loads of animals and machinery and specials of imported cattle moved about from the ports to the fattening areas. Periods of military activity produced trainloads of tanks, guns and suchlike; even target trains. Rebalancing motive power could result in the movement of three of four locomotives coupled together and the insatiable demands of a power station led to a merry-go-round load of wagons which discharged while on the move. Publicity trains, oil trains, cement trains, all added to the variety.

A sad sight occasionally appearing in wartime might be an ambulance train, an unusual but very efficient means of transporting injured servicemen for hospitalisation. The leading vehicle would be a brake van with part in use for infectious cases, and then with a staff coach and a kitchen coach following. Next would be nine ward coaches with a pharmacy strategically placed in the centre, while the remaining three vehicles would be used for a second kitchen, for staff purposes and one for stores. Other facilities would include fan ventilation, bathrooms, a huge water supply and, of course, lighting and heating.

DESIGN FRENZY

The basic railway operation using flanged-wheeled vehicles on iron rails and hauled by a steam-powered machine did not emerge unchallenged. Horsepower succumbed in the process early despite a brave showing by *Cycloped* at the Rainhill trials [see p. 68]. Stationary engines and rope/wire haulage did somewhat better and gave a good account of themselves until conventional steam locomotives became more powerful. Indeed the cable-operated London & Blackwall Railway used this system successfully from 1840 until locomotive haulage took over in 1849.

The so-called 'atmospheric railway' also made a strong challenge in the early years with successful operation on the London & Croydon line for a short time and major support from Brunel in terms of the South Devon Railway. The great man also had it in mind for a line from Bristol out to the Severn Estuary at Portishead, where a dock was to be built

Mr. Parkins's motive apparatus showing also the small guiding wheels and the winch for giving by hand slow motion to the train.

for international sailings, but the crucial leather pipes could not be kept airtight and conventional locomotives again triumphed.

Byways, in the search for other forms of motive power, were explored by several early engineers. Mr Kollman's trains used wheels with no flanges but relied on horizontal wheels acting on a central rail. Major Parley's so-called 'safety railway carriages' also embodied flangeless wheels but were kept stable by smaller flangeless wheels acting horizontally on the main track. Gradually, the rival ideas faded away but left behind some interesting drawings in the *Illustrated London News* (like the one above) and the archives of the Patent Office.

RACING

Rivalry between individual railway concerns resulted in some spectacular train performances over the years. Notable examples included the competition between the Southern and Great Western interests for the Exeter and Plymouth traffic and that of the London, Midland & Scottish Railway (LMS) and London & North Eastern Railway (LNER) companies and constituents over the East Coast and West Coast main lines northwards to Scotland. In addition to the keen express running these produced, the revenue lure of the major urban conurbations was also the subject of tight train timings and some very smart running.

More locally, enginemen frequently had their own minor contests, these liable to occur wherever two rival routes paralleled one another. An LMS locomotive and crew heading south from Gloucester found it hard to resist taking on the GWR local service on its way to Stroud and Swindon as the two ran side by side as far as Standish Junction. The same situation occurred on the parallel routes north from Peterborough where an LMS train for Stamford and beyond would readily take on one running on the Great Northern Railway (GNR) main line until it had to stop at Helpston station. And even at Kings Cross a cheeky set of suburban articulated coaches and its 0-6-0T locomotive would often heave its way up and out of a suburban station platform to take on a Pacific-headed main line express that was struggling to get hold of its heavy train on the adjacent main line.

Nowhere, though, was train working rivalry more strong than in the competition between the Glasgow & South Western (G&SWR) and Caledonian companies for the passengers heading to and from the Clyde coast piers and the steamers serving Dunoon and Rothesay on the Isle of Bute. At one time, the 'Caley' ran a non-stop train over the 26½ miles from Glasgow Central to Gourock in just thirty-two minutes, despite the route constraints at Paisley and Port Glasgow and a series of curves on the approach to Gourock Pier. The G&SWR response was to offer a thirty-four minute run over its 1 mile shorter line from St Enoch to Greenock.

This, too, was spectacular owing to the steep climb beyond Houston and the drop down to Cartsburn Tunnel.

These two competing trains arrived at their respective pier stations within three minutes of one another. Passengers were hustled along to the waiting rival steamers, which then continued the race across to Dunoon, sometimes with some questionable tactics being employed to be the first to get the piermaster's authority to moor. All great fun for some, but perhaps not for the nervous!

THE RACER, THE RASHER AND THE SPUD

The London & North Eastern Railway named a few of its freight trains, not very imaginatively, in a little booklet entitled *How the LNER Expresses Freight*. More recent years were to offer such examples as *The Lea Valley Enterprise* north from Waltham Cross and *The Clayliner* from Cornwall to the potteries. However, pride of place in the naming stakes was undoubtedly earned by the Great Western with its freight publicity booklet *How to Send and How to Save*. This listed no less than seventy-five names for its freight trains, many of them recognising informal titles the staff had been using for ages.

Three departures from Acton were *The High Flyer*, *The Early Bird* and *The Leek* and from Banbury a 3.40 a.m. train to Bristol was *The Competitor*. The 9.35 p.m. from Basingstoke to Wolverhampton was *The BBC*, denoting Basingstoke–Birmingham–Crewe as its function. Five trains from Birkenhead were named, but quite prosaically, like *The Pedlar* from Birmingham to Paddington or *The Hardware* from Bordesley Junction to Swansea.

Seven goods workings from Bristol were labelled, including *The Bacca*, *The Cocoa* and *The Drake*, and seven out of South Wales including *The Stock*, *The Spud* and *The Tinman*, which was the 10.23 p.m. from the steelworks at Margam heading for the distribution yards via Bordesley. In the reverse direction *The Bacon* left Gloucester at 12.05 a.m. for Cardiff while, unsurprisingly, the 8.20 p.m. Kidderminster to Paddington was called *The Carpet.*

The Great Western's publicity was without equal among the railway companies. In addition to the conventional poster and other tools it used such items as jigsaws and 'Legend Land' leaflets to promote passenger travel, while these freight train names were a clever combination of traditional staff nomenclature and titles imaginatively drawing attention to a freight service and its role. The numerous freight services out of Paddington for example included, predictably, *The Sauce* leaving for Worcester at 12.05 a.m. and the following 12.30 a.m. to Bristol, *The Mopper Up*. The long haul up from Penzance to Paddington was carried out by the 2.50 p.m., *The Searchlight* and the evening short run to Plymouth by *The Pasty*.

This period was one in which food production for the trade was a major part of the GWR freight business. So it was that the 10.30 p.m. Reading to Laira was named, entirely predictably, *The Biscuits*, as its main load emanated from Huntley & Palmer's private siding. *The Grocer* started at Southall and *The Rasher* at Swindon and Westbury despatched the 4.20 a.m. *The Moonraker* to Wolverhampton, *The Lancashire Lad* to Manchester and *The Western Flash* to Penzance. Some of the other freight train names were more prosaic, but not *The Racer*, *The Crosser* and *The Flying Skipper* out of Wolverhampton. Nor, reflecting the fertile county of Worcestershire, were the Worcester trains: *The Worcester Fruit* and *The 'Sparagras.*

If you enjoyed this book, you may also be interested in…

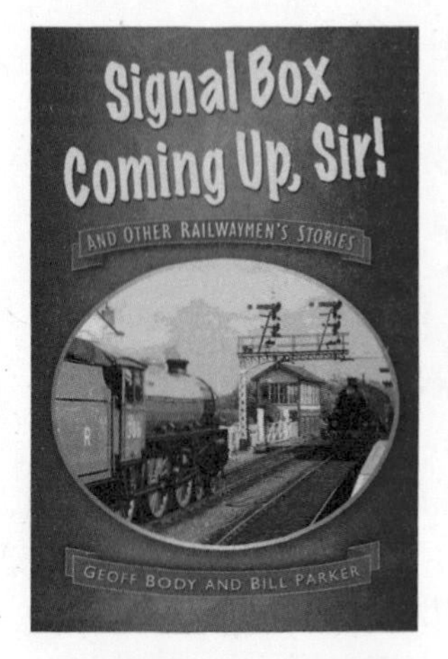

Signal Box Coming Up, Sir!

GEOFF BODY AND BILL PARKER

There's never a dull moment in this entertaining collection of experiences as Geoff Body and Bill Parker present hilarious highlights from the careers of railwaymen around Britain over the last fifty years. Featuring daring robberies, royal visits, lost passengers, bomb scares, coffins, circus trains and ladies of the night, it chronicles both successes and disasters, with accounts of moving a farm and a circus, 245 miles of marooned railway, footplate adventures, animal capers and many equally fascinating subjects.

978 0 7524 6040 6

Real Railway Tales

GEOFF BODY AND BILL PARKER

Running a railway is a complex business, constantly throwing up drama, misadventure and the unexpected. Geoff Body and Bill Parker have collated a rich selection of railwaymen's memories and anecdotes to create an enjoyable book of escapades and mishaps, illustrating the daily obstacles faced on the railways, from handling the new Eurostar to train catering, nights on the Tay Bridge to rail 'traffic cops', and from mystery derailments to track subsidence.

978 0 7509 5635 2

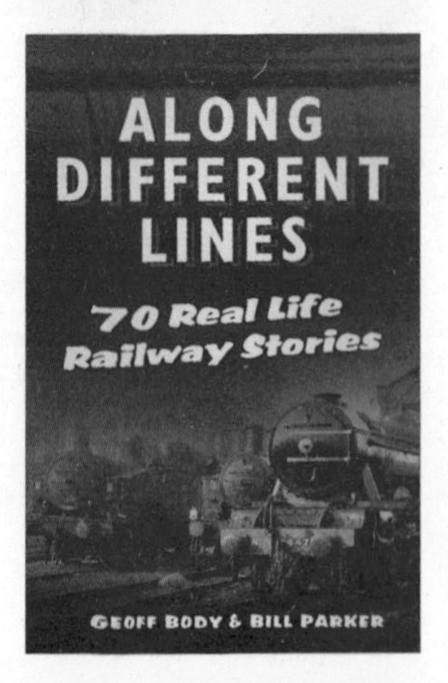

Along Different Lines

GEOFF BODY AND BILL PARKER

The incidents covered in this illustrated book include such bizarre 'everyday' events as coping with hurricanes, rogue locomotives and runaway wagons. There are PR successes and flops, the *Brighton Belle*, *Flying Scotsman* and *Mallard*, training course capers, a wino invasion, trackside antics and memorable royal visits. However well organised, the operation of a railway will always have its surprises; often hilarious, frequently unexpected, and sometimes serious. Two railway professionals recall notable incidents in this enjoyable look back at real life on the railways.

978 0 7524 8915 5

Visit our website and discover thousands of other History Press books.

www.thehistorypress.co.uk